Springer Series in Statistics

Probability and its Applications

A Series of the Applied Probability Trust

Springer Series in Statistics

Anderson: Continuous-Time Markov Chains: An Applications-Oriented Approach.
Andrews/Herzberg: Data: A Collection of Problems from Many Fields for the Student and Research Worker.
Anscombe: Computing in Statistical Science through APL.
Berger: Statistical Decision Theory and Bayesian Analysis, 2nd edition.
Bolfarine/Zacks: Prediction Theory for Finite Populations.
Brémaud: Point Processes and Queues: Martingale Dynamics.
Brockwell/Davis: Time Series: Theory and Methods, 2nd edition.
Choi: ARMA Model Identification
Daley/Vere-Jones: An Introduction to the Theory of Point Processes.
Dzhaparidze: Parameter Estimation and Hypothesis Testing in Spectral Analysis of Stationary Time Series.
Farrell: Multivariate Calculation.
Fienberg/Hoaglin/Kruskal/Tanur (Eds.): A Statistical Model: Frederick Mosteller's Contributions to Statistics, Science, and Public Policy.
Goodman/Kruskal: Measures of Association for Cross Classifications.
Grandell: Aspects of Risk Theory.
Hall: The Bootstrap and Edgeworth Expansion.
Härdle: Smoothing Techniques: With Implementation in S.
Hartigan: Bayes Theory.
Heyer: Theory of Statistical Experiments.
Jolliffe: Principal Component Analysis.
Kotz/Johnson (Eds.): Breakthroughs in Statistics Volume I.
Kotz/Johnson (Eds.): Breakthroughs in Statistics Volume II.
Kres: Statistical Tables for Multivariate Analysis.
Leadbetter/Lindgren/Rootzén: Extremes and Related Properties of Random Sequences and Processes.
Le Cam: Asymptotic Methods in Statistical Decision Theory.
Le Cam/Yang: Asymptotics in Statistics: Some Basic Concepts.
Manoukian: Modern Concepts and Theorems of Mathematical Statistics.
Miller, Jr.: Simultaneous Statistical Inference, 2nd edition.
Mosteller/Wallace: Applied Bayesian and Classical Inference: The Case of *The Federalist Papers.*
Pollard: Convergence of Stochastic Processes.
Pratt/Gibbons: Concepts of Nonparametric Theory.
Read/Cressie: Goodness-of-Fit Statistics for Discrete Multivariate Data.
Reiss: Approximate Distributions of Order Statistics: With Applications to Nonparametric Statistics.
Ross: Nonlinear Estimation.
Sachs: Applied Statistics: A Handbook of Techniques, 2nd edition.
Salsburg: The Use of Restricted Significance Tests in Clinical Trials.
Särndal/Swensson/Wretman: Model Assisted Survey Sampling.
Seneta: Non-Negative Matrices and Markov Chains.

(continued after index)

Much of human experience is random,
and almost all of it unexpected ....

J. Gani
Journal of Applied Probability
Special Volume 25A (1988), p. 3.

Preface

During the last two decades, considerable progress has been made in statistical time series analysis. In particular, numerous techniques utilizing the autoregressive moving-average (ARMA) model have been developed. Three books have played important roles in these developments; they are the texts by Box and Jenkins (1976), Hannan (1970a), and Anderson (1971).

ARMA modeling is usually classified into three stages. The first, model identification, aims to determine the orders of the ARMA model. The second, model estimation, is designed to estimate the ARMA coefficients. The last, diagnostic checking, examines the goodness-of-fit of the estimated model. Since the above-mentioned books appeared, various new methods have been proposed for the three stages, particularly for identification. Previously, methods based on the testing of hypotheses had been popularly employed as ARMA model identification procedures. In their book, Box and Jenkins (1976) used the autocorrelation function and the partial autocorrelation function. In the early 1970s, Akaike presented two penalty function identification methods known as the FPE and the AIC. To circumvent the inconsistency problem of the criteria, the BIC and Hannan and Quinn's method were proposed. At the same time, Cleveland suggested the use of the inverse autocorrelation function. In 1982, Hannan and Rissanen employed the instrumental regression technique as well as the penalty functions for ARMA modeling. This method has resulted in consistent estimates of the orders. Similar methods have been proposed by Koreisha and Pukkila. Throughout the 1980s, identification methods using the patterns of some functions of the autocorrelations have been studied. These are called pattern identification methods and include Woodside's method, the R and S method of Gray, Kelley, and McIntire, the Corner method of Beguin, Gourieroux and Monfort, the three GPAC methods of Woodward and Gray, of Glasbey, and of Takemura, the ESACF method of Tsay and Tiao, the SCAN method of Tsay and Tiao, and the 3-pattern method of Choi.

Although some of the identification methods are mentioned in the literature, there is no book that brings all of them together. The aim of this book is to give an account of all the identification methods mentioned above. Of course, it is impossible to examine all of these in detail. However, I have tried to survey as many identification methods as possible. Because my purpose is to explain all the methods in this short book, what I have emphasized is not the mathematical details of the methods but rather their fundamental ideas. Appropriate references are provided for further illustrative explorations and mathematical investigations. At the end of each chapter, additional references are given; these are basically for graduate students starting research in time series analysis, and for readers who want to apply recently developed time series analysis techniques to their research. Although this book may at first sight appear to be too difficult for graduate students, I hope that it can be used as an auxiliary textbook for graduate courses. It is, in fact, based on my lecture notes for a time series analysis course taught at Yonsei University in Seoul.

This monograph was written during my sabbatical leave at the Department of Statistics at the University of California, Santa Barbara (UCSB). I must thank the faculty, staff, and graduate students for their friendliness and help during my stay; I wish to express my gratitude to Professor Joe Gani, who has led me to understand more deeply the values of human and scholarly life. His warmth and consideration have helped to ease the burden of the death of one very close to me in 1990.

I am greatly indebted to H. Akaike, E. J. Hannan, J. R. M. Hosking, S. G. Koreisha, E. Parzen, T. M. Pukkila, J. Rissanen, G. C. Tiao, and R. S. Tsay, whose comments on an early version of the manuscript led to substantial improvements. My friends Seong-Cheol Cho, KiYoung Choi, MooYoung Choi, Hay Y. Chung, Myung-Hoe Huh, SeJung Oh, and Sukgoo Pak read drafts carefully, and I appreciate their contributions. I would also like to express my gratitude to current and former graduate students Wonkyung Lee, Hyun-Ju Noh, GoungJoo Shin, and SeongBack Yi at Yonsei University and to Benny Cheng, Jie Huang, Aaron Gross, and Jeffrey Stein at UCSB as well as to Mr. F. Schubert at the UCSB Humanities Computing Facility. Finally, I wish to thank my friend HoYoun Kim for his support.

CBS

Contents

1

Introduction

1.1 ARMA Model

Consider the autoregressive moving-average (ARMA) model of orders p and q,

$$\phi(B)y_t = \theta(B)v_t, \tag{1.1}$$

where $\phi(B) = -\phi_0 - \phi_1 B - \cdots - \phi_p B^p$, $\theta(B) = -\theta_0 - \theta_1 B - \cdots - \theta_q B^q$, $\phi_0 = \theta_0 = -1$, $\phi_p \neq 0$, $\theta_q \neq 0$, B is the backshift operator, and $\{v_t\}$ is a sequence of independent and identically distributed random variables with means 0 and variances σ^2 (> 0). The sequence $\{v_t\}$ is called either a white noise process or an innovation process. In some time series books, the white noise process is defined as a sequence of uncorrelated random variables instead of that of independent random variables. In practical time series analysis, there is not as much difference between the two definitions. We assume that the model is stationary and invertible, i.e., the equations $\phi(z) = 0$ and $\theta(z) = 0$ have all the roots outside the unit circle. We assume that the two equations have no common root. This assumption is sometimes called coprimal. The stationarity and the invertibility conditions have been discussed by several authors. Interested readers may consult the references in Section 1.6. In statistical literature, the white noise process is frequently assumed to be Gaussian, i.e., normally distributed. There are some references about non-Gaussian ARMA processes in Section 1.6. In this book, we assume that the coefficients $\phi_1, \ldots, \phi_p$, $\theta_1, \ldots, \theta_q$, and the white noise variance σ^2 are constants, i.e., they do not depend on time. There is some statistical literature about ARMA models with time-varying coefficients and with random coefficients. Interested readers may consult the references in Section 1.6.

Actually, we can barely anticipate that a T-realization $\{y_1, \ldots, y_T\}$ is from an exact ARMA(p, q) model. We may regard the ARMA model as a simplified alternative to the complicated nature. It is well-known that a stationary stochastic process can be approximated by a suitable ARMA model. (See, e.g., Doob [1953, pp. 498-506].) Henceforth, we assume the underlying process is from an exact ARMA(p, q) model. If the orders p and q are predetermined, then the parameters $\phi_1, \ldots, \phi_p$, $\theta_1, \ldots, \theta_q$, and σ^2 can be estimated from the realization by some estimation methods. However, because the true orders are not known a priori, we should determine the orders based on the observations. This procedure has been named the *model identification* by Box and Jenkins (1976), which is a challenging but distressing problem in ARMA modeling. The purpose of this book is to explain appropriate ARMA model identification methods. Even though it is impossible to examine all identification methods in detail, we will try to survey as many as possible. We will not emphasize mathematical details of each method but its fundamental ideas. For further illustrative examination or mathematical exploration, pertinent references will be provided.

Because the process is assumed to be stationary, the autocovariance function (ACVF) and the autocorrelation function (ACRF) can be defined as

$$\begin{aligned} \sigma(j) &= \operatorname{cov}(y_t, y_{t+j}), \quad j = 0, \pm 1, \pm 2, \ldots, \\ \rho_j &= \sigma(j)/\sigma(0), \quad j = 0, \pm 1, \pm 2, \ldots. \end{aligned}$$

It is well-known (see, e.g., Priestley [1981, p. 140]) that the autoregressive (AR) parameters satisfy the extended Yule-Walker (EYW) equations

$$\sigma(j) = \phi_1 \sigma(j-1) + \cdots + \phi_p \sigma(j-p), \quad j = q+1, q+2, \ldots$$

or, equivalently,

$$\rho_j = \phi_1 \rho_{j-1} + \cdots + \phi_p \rho_{j-p}, \quad j = q+1, q+2, \ldots.$$

It is also known (Choi [1986a]) that the parameters and the ACVF satisfy the relations

$$\begin{aligned} &\phi_0 \sigma(j) + \phi_1 \sigma(j-1) + \cdots + \phi_p \sigma(j-p) \\ &\qquad = (\psi_0 \theta_j + \psi_1 \theta_{j+1} + \cdots + \psi_{q-j} \theta_q) \sigma^2, \quad j = 0, \ldots, q. \end{aligned} \tag{1.2}$$

Here ψ_j is defined by

$$\phi_0 \psi_j + \phi_1 \psi_{j-1} + \cdots + \phi_j \psi_0 = \begin{cases} \theta_j, & j = 0, \ldots, q, \\ 0, & j = q+1, q+2, \ldots, \end{cases}$$

where ϕ_j is assumed to be 0 for $j = p+1, p+2, \ldots$. If we let

$$\psi(z) = \sum_{l=0}^{\infty} \psi_l z^l,$$

then $\phi(z)\psi(z) = \theta(z)$.

1.2 History

If $p = 0$, then the ARMA process is called a pure moving-average (MA) process, which was first introduced by Yule (1921) and Slutzky (1927, 1937). Yule (1927) introduced a pure autoregressive process, which corresponds to $q = 0$. As far as the present author knows, a mixed ARMA process was first used by Walker (1950). There are some references to the history of using the ARMA model in Section 1.6.

There have been a lot of advances in time series analysis during the last two decades. Several papers summarize recent developments like the series papers by Newbold (1981, 1984, 1988). For more survey papers, readers may refer to Section 1.6. There are also many books of collected papers of time series analysis. Particularly, two volumes in the *Handbook of Statistics* series cover recent progresses in time series analysis well. One is *Time Series in the Frequency Domain (Vol. 3)* edited by Brillinger and Krishnaiah (1983), and the other is *Time Series in the Time Domain (Vol. 5)* edited by Hannan, Krishnaiah, and Rao (1985). For more books of collected papers, refer to Section 1.6.

Some review papers of time series model identification have been presented in engineering fields by Åström and Eykhoff (1971), Unbehauen and Göhring (1974), and Van den Boom and Van den Enden (1974). Anděl (1982) reviewed most of the ARMA identification methods which had appeared in statistical literature by that time. Piccolo and Tunnicliffe-Wilson (1984) reviewed several identification methods for ARMA processes and tried to unify them through a covariance matrix. Shibata (1985), Stoica, Eykhoff, Janssen and Söderström (1986), and Koreisha and Yoshimoto (1991) provided brief reviews of ARMA model identification. Gooijer, Abraham, Gould, and Robinson (1985) presented a detailed review. Lütkepohl (1985) compared various criteria for determining the orders of vector AR processes in a simulation study. In the paper, his main concern was the performance of various identification methods in small samples.

1.3 Algorithms

When the ACVF of the ARMA(p, q) process (1.1) is given, the AR parameters, the MA parameters and the white noise variance can be obtained by solving the EYW equations and Equation (1.2). In other words, we can obtain the $p+q+1$ parameters $\phi_1, \ldots, \phi_p$, $\theta_1, \ldots, \theta_q$, and σ^2 from the first $p+q+1$ autocovariance terms $\sigma(0), \ldots, \sigma(p+q)$ by solving the equations. Because the AR part of the equations is linear, it can be easily solved. In contrast, the MA part is composed of nonlinear equations. Thus, some iterative methods are needed to solve it.

1.3.1 AR Parameters

Because the EYW equations are of linear form, we may use the Gauss elimination method to solve them. To identify an ARMA process, it is necessary to solve the EYW equations for several pairs of orders. Therefore, the Gauss elimination method is not so practical.

We define $\phi_{k,1}^{(i)}, \ldots, \phi_{k,k}^{(i)}$ as the solutions of the following EYW equations:

$$\rho_j = \phi_{k,1}^{(i)}\rho_{j-1} + \phi_{k,2}^{(i)}\rho_{j-2} + \cdots + \phi_{k,k}^{(i)}\rho_{j-k}, \quad j = i+1, \ldots, i+k,$$

where the subscript k and the superscript (i) mean the AR order and the MA order, respectively. Particularly, if $i = 0$, then the superscript (0) will be omitted. We define k-dimensional Toeplitz matrices and some vectors by

$$\Sigma(k,i) = \begin{pmatrix} \sigma(i) & \sigma(i-1) & \cdots & \sigma(i-k+1) \\ \sigma(i+1) & \sigma(i) & \cdots & \sigma(i-k+2) \\ \vdots & \vdots & & \vdots \\ \sigma(i+k-1) & \sigma(i+k-2) & \cdots & \sigma(i) \end{pmatrix},$$

$$B(k,i) = \frac{1}{\sigma(0)}\Sigma(k,i),$$

$$\boldsymbol{\sigma}(k,i) = (\sigma(i+1), \ldots, \sigma(i+k))^t,$$

$$\boldsymbol{\phi} = (\phi_1, \ldots, \phi_p)^t,$$

$$\boldsymbol{\theta} = (\theta_1, \ldots, \theta_q)^t,$$

$$\boldsymbol{\phi}_* = (\phi_0, \phi_1, \ldots, \phi_p)^t,$$

$$\boldsymbol{\theta}_* = (\theta_0, \theta_1, \ldots, \theta_q)^t,$$

$$\boldsymbol{\phi}(k,i) = (\phi_{k,1}^{(i)}, \ldots, \phi_{k,k}^{(i)})^t.$$

With the above notations, the EYW equations can be represented by

$$\Sigma(k,i)\boldsymbol{\phi}(k,i) = \boldsymbol{\sigma}(k,i).$$

If $\{y_t\}$ is from the ARMA(p,q) model (1.1), then the EYW equations imply that $\Sigma(k,i)$ is nonsingular for $(k,i) \in \{(k,q) \mid k = p, p+1, \ldots\} \bigcup \{(p,i) \mid i = q, q+1, \ldots\}$. Thus, if $k \geq p$, then

$$\boldsymbol{\phi}(k,q) = (\phi_1, \ldots, \phi_p, 0, \ldots, 0)^t.$$

If $k = p+1, p+2, \ldots$ and $i = q+1, q+2, \ldots$, then $\Sigma(k,i)$ is singular, and then the solutions of $\Sigma(k,i)\boldsymbol{\phi}(k,i) = \boldsymbol{\sigma}(k,i)$ constitute a subspace of R^k spanned by $\{(\phi_1, \ldots, \phi_p, 0, \ldots, 0), (0, \phi_1, \ldots, \phi_p, 0, \ldots, 0), \ldots, (0, \ldots, 0, \phi_1, \ldots, \phi_p)\}$.

We consider pure AR processes. In this case the EYW equations become the Yule-Walker (YW) equations, and $\{\phi_{k,1}, \ldots, \phi_{k,k} \mid k = 1, 2, \ldots\}$ can be calculated by the famous Levinson (1947)-Durbin (1960a) algorithm.

Algorithm 1.1. *The Levinson-Durbin Algorithm*

For $k = 0$, let

$$\phi_{1,1} = \rho_1,$$
$$\lambda(1) = 1 - \phi_{1,1}^2.$$

For $k = 1, 2, \ldots,$

$$\theta(k) = \rho_{k+1} - \phi_{k,1}\rho_k - \cdots - \phi_{k,k}\rho_1,$$

$$\phi_{k+1,k+1} = \frac{\theta(k)}{\lambda(k)},$$

$$\lambda(k+1) = \lambda(k)\{1 - \phi_{k+1,k+1}^2\}.$$

For $j = 1, 2, \ldots, k,$

$$\phi_{k+1,j} = \phi_{k,j} - \phi_{k+1,k+1}\phi_{k,k+1-j}. \quad \square$$

The white noise variance of the AR process satisfies

$$\sigma^2 = \sigma(0) - \phi_1\sigma(1) - \cdots - \phi_p\sigma(p).$$

If we define

$$\sigma_k^2 = \sigma(0) - \phi_{k,1}\sigma(1) - \cdots - \phi_{k,k}\sigma(k), \quad k = 1, 2, \ldots,$$

then

$$\sigma_k^2 = \sigma(0)\lambda(k), \quad k = 1, 2, \ldots.$$

As Akaike (1973b) mentioned, the Levinson-Durbin algorithm is probably one of the most significant contributions in the field of digital computer analysis of time series. Its numerical stability was questioned by Pagano (1972) and Box and Jenkins (1976, p. 84). Cybenko (1979, 1980) has shown that

$$\frac{1}{\min\left\{\prod_{j=1}^{k-1}(1-\phi_{j,j}^2), \prod_{j=1}^{k-1}(1-\phi_{j,j})\right\}} \leq \| B^{-1}(k,0) \| \leq \prod_{j=1}^{k-1} \frac{1+ \mid \phi_{j,j} \mid}{1- \mid \phi_{j,j} \mid},$$

where $\| A \|$ is a norm of matrix A. From these inequalities we know that the numerical instability is not due to the algorithm but due to the nature of the covariance matrix $\Sigma(k,i)$. The inequalities show that we should not rely on the solutions of the YW equations if any root of $\phi(z) = 0$ is near the unit circle.

Tong (1988) has interpreted the Levinson-Durbin algorithm using a local parameter orthogonality of $\phi_{p+1,p+1}$ and $(\phi_{p,1}, \ldots, \phi_{p,p})$. Consider the reparameterization of $\boldsymbol{\phi}(p+1,0)$ to $(\boldsymbol{\phi}(p,0)^t, \phi_{p+1,p+1})$ by

$$\phi_{p+1,j} = \phi_{p+1,j}(\boldsymbol{\phi}(p,0)^t, \phi_{p+1,p+1}), \quad j = 1, \ldots, p.$$

The purpose here is to reparameterize them so that $\phi_{p+1,p+1}$ and $(\phi_{p,1}, \ldots, \phi_{p,p})$ are orthogonal in Cox-Reid's sense (1987). Under the normality assumption, Cox-Reid's orthogonality condition yields

$$\sum_{l=1}^{p} \rho_{j-l} \frac{\partial \phi_{p+1,l}}{\partial \phi_{p+1,p+1}} = \rho_{p+1-j}, \quad j = 1, \ldots, p.$$

Because the YW equations are

$$\sum_{l=1}^{p} \rho_{j-l} \phi_{p,p+1-l} = \rho_{p+1-j}, \quad j = 1, \ldots, p$$

and because $B(p, 0)$ is nonsingular,

$$\frac{\partial \phi_{p+1,j}}{\partial \phi_{p+1,p+1}} = \phi_{p,p+1-j}, \quad j = 1, \ldots, p.$$

Thus, the condition of the locally orthogonal reparameterization of $\boldsymbol{\phi}(p+1, 0)$ is

$$\phi_{p+1,j} = \phi_{p+1,p+1}\phi_{p,p+1-j} + c_j(\boldsymbol{\phi}(p, 0)), \quad j = 1, \ldots, p,$$

where $c_j(\boldsymbol{\phi}(p, 0))$ is a function of $\boldsymbol{\phi}(p, 0)$. As an example satisfying the condition, consider

$$\phi_{p+1,j} = \phi_{p,j} + \phi_{p+1,p+1}\phi_{p,p+1-j}, \quad j = 1, \ldots, p,$$

which is the Levinson-Durbin algorithm. Based on the orthogonality, Tong has mentioned that the YW estimates $\hat{\phi}_{p+1,p+1}$ and $\hat{\boldsymbol{\phi}}(p, 0)$ obtained through the Levinson-Durbin algorithm are asymptotically independent when the true order of the AR model is $p + 1$. It can be easily guessed, for $\phi_{p+1,p+1}$ is the partial autocorrelation of y_t and y_{t+p+1} given $y_{t+1}, \ldots, y_{t+p}$. Morettin (1984) presented a detailed review of the Levinson-Durbin algorithm.

Consider mixed ARMA processes. When the MA order is fixed, the system of the EYW equations constitutes a nested Toeplitz system. Thus, we may use the Trench(1964)-Zohar(1969, 1974, 1979), the Berlekamp(1968)-Massey(1969), or the recursive Euclidean algorithms (see, e.g., Blahut [1985]). Among them we are going to state the Trench-Zohar algorithm. We consider the case $i = q$, i.e., the MA order is q.

Algorithm 1.2. *A Simplified Trench-Zohar Algorithm*
Initial values for recursion:

$$\theta(0, q) = \rho_{q+1},$$
$$\eta(0, q) = \rho_{q-1},$$

$$\lambda(0,q) = \rho_q,$$
$$\phi_{1,1}^{(q)} = \frac{\theta(0,q)}{\lambda(0,q)},$$
$$\pi_{1,1}^{(q)} = \frac{\eta(0,q)}{\lambda(0,q)},$$

$$\lambda(1,q) = \lambda(0,q)\left\{1 - \phi_{1,1}^{(q)}\pi_{1,1}^{(q)}\right\}.$$

For $k = 1, 2, \ldots,$

$$\theta(k,q) = \rho_{q+k+1} - \phi_{k,1}^{(q)}\rho_{q+k} - \cdots - \phi_{k,k}^{(q)}\rho_{q+1},$$
$$\eta(k,q) = \rho_{q-k-1} - \pi_{k,1}^{(q)}\rho_{q-k} - \cdots - \pi_{k,k}^{(q)}\rho_{q-1},$$

$$\phi_{k+1,k+1}^{(q)} = \frac{\theta(k,q)}{\lambda(k,q)},$$
$$\pi_{k+1,k+1}^{(q)} = \frac{\eta(k,q)}{\lambda(k,q)},$$

$$\lambda(k+1,q) = \lambda(k,q)\{1 - \phi_{k+1,k+1}^{(q)}\pi_{k+1,k+1}^{(q)}\}.$$

For $j = 1, \ldots, k,$

$$\phi_{k+1,j}^{(q)} = \phi_{k,j}^{(q)} - \phi_{k+1,k+1}^{(q)}\pi_{k,k+1-j}^{(q)},$$
$$\pi_{k+1,j}^{(q)} = \pi_{k,j}^{(q)} - \pi_{k+1,k+1}^{(q)}\phi_{k,k+1-j}^{(q)}. \quad \square$$

Algorithm 1.2 is a simplified version of the Trench-Zohar algorithm about Toeplitz matrix inversion and is also a generalized version of the Levinson-Durbin algorithm. There are more references about Algorithm 1.2 in Section 1.6.

The following algorithms are useful when neither the AR order nor the MA order is known (Choi [1991i]).

Algorithm 1.3. *A Generalized Levinson-Durbin Algorithm*
For $i = 0$, use the Levinson-Durbin Algorithm to calculate

$$\{\phi_{k,j}^{(0)} \mid k = 1, 2, \ldots, \quad j = 1, \ldots, k\}.$$

For $i = 1, 2, \ldots,$

For $k = 0$, let

$$\lambda(0,i) = \rho_i,$$
$$\phi_{1,1}^{(i)} = \frac{\rho_{i+1}}{\rho_i}.$$

For $k = 1, 2 \ldots,$

$$\theta(k,i) = \rho_{i+k+1} - \phi_{k,1}^{(i)} \rho_{i+k} - \cdots - \phi_{k,k}^{(i)} \rho_{i+1},$$

$$\lambda(k,i) = \lambda(k-1,i) \left\{ 1 - \frac{\phi_{k,k}^{(i)}}{\phi_{k,k}^{(i-1)}} \right\},$$

$$\phi_{k+1,k+1}^{(i)} = \frac{\theta(k,i)}{\lambda(k,i)}.$$

For $j = 1, \ldots, k,$

$$\phi_{k+1,j}^{(i)} = \phi_{k,j}^{(i)} + \phi_{k+1,k+1}^{(i)} \frac{\phi_{k,j-1}^{(i-1)}}{\phi_{k,k}^{(i-1)}}. \quad \square$$

We can derive the above algorithm using the bordering technique of matrix inversion (see, e.g., Faddeeva [1959, pp. 107-111]) and the property that the inverse of a persymmetric matrix is also persymmetric. Clearly, Algorithm 1.3 is another generalization of the Levinson-Durbin algorithm to mixed ARMA processes. It is computationally more efficient than any other existing one to solve the EYW equations for AR parameters, particularly when the true orders are unknown. Moreover, it does not require one to calculate the dummy sequences $\{\pi_{k,j}^{(i)}\}$.

Choi (1990d) has derived the following algorithm, which is closely related to Algorithm 1.3.

Algorithm 1.4. *A Generalized Levinson-Durbin Algorithm*
For $i = 0$, use the Levinson-Durbin algorithm to calculate

$$\{\phi_{k,j}^{(0)} \mid k = 1, 2, \ldots, \; j = 1, \ldots, k\}.$$

For $i = 1, 2, \ldots,$

For $k = 1, 2 \ldots,$

$$\phi_{k,0}^{(i)} = -1$$

For $j = 1, \ldots, k,$

$$\phi_{k,j}^{(i)} = \phi_{k+1,j}^{(i-1)} - \frac{\phi_{k+1,k+1}^{(i-1)}}{\phi_{k,k}^{(i-1)}} \phi_{k,j-1}^{(i-1)}. \quad \square$$

Algorithm 1.4 is the same as Tsay and Tiao's algorithm (1984) for calculating the extended sample autocorrelation function (ESACF) under the stationarity assumption. Thus, the two algorithms will produce the same estimates as long as they are initialized in the same way. Also, refer to Pham (1988).

1.3.2 MA Parameters

Theoretically, we can obtain the MA parameters and the white noise variance by solving the simultaneous equations in (1.2) when the ACVF is given. Because they are highly nonlinear, some iterative methods are required to solve them. Also, it leaves the uniqueness problem of the solution.

To consider the uniqueness problem, we define the covariance generating function by

$$g(z) = \sum_{j=-\infty}^{\infty} \sigma(j) z^j.$$

Then, we get

$$\phi(z)\phi(z^{-1})g(z) = \sigma^2\theta(z)\theta(z^{-1}).$$

Denote its RHS by $A(z)$. If $z_1, \ldots, z_q$ denote the reciprocals of the roots of $\theta(z) = 0$, then $A(z)$ can be written as

$$A(z) = \sigma^2 \prod_{j=1}^{q} (1 - z_j z)(1 - z_j z^{-1}).$$

If any real value of z_j is replaced by z_j^{-1}, or if any pair $(z_j, \bar{z}_j)$ of complex conjugate values are by their reciprocals, then $A(z)$ remains unaltered up to σ^2. Therefore, any different form of

$$\theta(z) = \prod_{j=1}^{q} (1 - w_j z),$$

where w_j is either z_j or z_j^{-1}, results in the same autocovariance structure. Consequently, the system of simultaneous equations in (1.2) has at most 2^q different solution sets for $\theta_1, \ldots, \theta_q$ and σ^2. To express y_t of the ARMA process with the past observation terms $y_{t-1}, y_{t-2}, \ldots$ and the current innovation term v_t, we usually impose the invertibility assumption on the ARMA process, i.e., $\theta(z) = 0$ has all the roots outside the unit circle. In this case, there is a unique solution set for $\theta_1, \ldots, \theta_q$ and σ^2. Throughout this book the invertibility will be assumed as mentioned before.

Choi (1986a) has developed an iterative algorithm to obtain the unique solution of the MA parameters and the white noise variance using a special property of triangular Toeplitz matrices. It is simple and computationally cheap. However, its convergence has not been proven yet.

Choi (1987) has presented a Newton-Raphson solution of the nonlinear simultaneous equations. Because the ARMA(p, q) process is stationary, it can be represented by the MA(∞) model

$$y_t = \psi(B) v_t,$$

where

$$\psi(B) = \phi^{-1}(B)\theta(B) = \sum_{j=0}^{\infty} \psi_j B^j.$$

Solving the simultaneous equations in (1.2) for $\theta_1, \ldots, \theta_q$ and σ^2 is equivalent to doing so for $\psi_1, \ldots, \psi_q$ and σ^2 when the AR parameters $\phi_1, \ldots, \phi_p$ are given.

Let $\Sigma(q+1, p+1; 0)$ and Ψ be $(q+1) \times (p+1)$ matrices, whose (i,j) elements are

$$(\Sigma(q+1,p+1;0))_{i,j} = \sigma(i-j),$$
$$(\Psi)_{i,j} = \begin{cases} 0, & i < j \\ \psi_{i-j}, & \text{otherwise,} \end{cases}$$

respectively. Also let Ψ_q be a $(q+1) \times (q+1)$ matrix whose (i,j) element is

$$(\Psi_q)_{i,j} = \begin{cases} 0, & i > j \\ \psi_{j-i}, & \text{otherwise.} \end{cases}$$

Then Equation (1.2) yields

$$\Sigma(q+1,p+1;0)\boldsymbol{\phi}_* = \sigma^2 \Psi_q \Psi \boldsymbol{\phi}_*.$$

Let $c_0 = \sigma, c_1 = \sigma\psi_1, \ldots, c_q = \sigma\psi_q$, $C_q = \sigma\Psi_q$, $C = \sigma\Psi$, and $\mathbf{c} = (c_0, c_1, \ldots, c_q)^t$. Then it becomes

$$C_q C \boldsymbol{\phi}_* = \Sigma(q+1,p+1;0)\boldsymbol{\phi}_*,$$

which we can solve by the following algorithm.

Algorithm 1.5. *A Newton-Raphson Algorithm*
For $j = 0, \ldots, q$, let T_j and S_j be $(q+1) \times (p+1)$ matrices, whose (r,s) elements are

$$(T_j)_{r,s} = \begin{cases} c_{j+r-s}, & 1 \le s \le j+r,\ 1 \le r \le q-j+1 \\ 0, & \text{otherwise,} \end{cases}$$
$$(S_j)_{r,s} = \begin{cases} c_{j-r+s}, & 1 \le r \le q+1,\ \max\{1, r-j\} \le s \le q-j+1, \\ 0, & \text{otherwise,} \end{cases}$$

respectively. Also, let

$$U_j = T_j + S_j, \quad j = 0, \ldots, q,$$
$$W = (U_0\boldsymbol{\phi}_*, \ldots, U_q\boldsymbol{\phi}_*).$$

Then the Newton-Raphson solution of $C_q C\boldsymbol{\phi}_* = \Sigma(q+1,p+1;0)\boldsymbol{\phi}_*$ is obtained by the recursive equation

$$\mathbf{c}^{(n+1)} = \frac{1}{2}\mathbf{c}^{(n)} + (W^{(n)})^{-1}\Sigma(q+1,p+1;0)\boldsymbol{\phi}_*,$$

where the superscript (n) means the value at the nth iteration. □

Wilson (1969) has derived the Newton-Raphson algorithm for a pure MA process, which is a special case of Algorithm 1.5. It is worth mentioning that Algorithm 1.5 shows Equation (4) of Wilson's paper can be simplified as

$$\boldsymbol{\theta}^{(t+1)} = \frac{1}{2}\boldsymbol{\theta}^{(t)} + (T^{(t)})^{-1}\mathbf{c}^*.$$

Choi (1987) has shown that Algorithm 1.5 has the second-order convergence, if proper starting values are used. To obtain approximate values of the MA parameters, we may use an MA(s) model with a fairly large s,

$$y_t^{(s)} = \sum_{j=0}^{s} \psi_j^{(s)} v_{t-j},$$

where $\psi_0^{(s)}, \ldots, \psi_s^{(s)}$ are to be determined so that $\operatorname{cov}(y_t^{(s)}, y_{t-j}^{(s)}) = \sigma(j)$ for $j = 0, \ldots, s$. It is known (see, e.g., T. W. Anderson [1971]) that $y_t^{(s)}$ converges to y_t in L^2 as $s \to \infty$. Thus, we can choose an integer s so that $\psi_0^{(s)}, \ldots, \psi_s^{(s)}$ are as close to $\psi_0, \ldots, \psi_s$ as we wish. If we apply Algorithm 1.5 to the MA(s) model with starting values

$$c_0^{(0)} = \{\sigma(0) + 2\sum_{j=1}^{s} \sigma(j)\}^{1/2},$$

$$c_j^{(0)} = \phi_1 c_{j-1}^{(0)} + \cdots + \phi_j c_0^{(0)}, \quad j = 1, \ldots, s,$$

where $\phi_{p+1} = \phi_{p+2} = \cdots = 0$, then the resulting values converge to the true parameter values satisfying the invertibility condition of the MA(s) model. Thus, they are suitable as starting values in applying Algorithm 1.5 to the ARMA(p,q) model. However, it takes a lot of computing time to obtain them when s is large. Choi (1987) has proposed another set of starting values

$$c_0^{(0)} = \{\sigma(0)\}^{1/2},$$

$$c_j^{(0)} = \phi_1 c_{j-1}^{(0)} + \cdots + \phi_j c_0^{(0)}, \quad j = 1, \ldots, q,$$

where $\phi_{p+1} = \phi_{p+2} = \cdots = 0$. These values are simply chosen so that $\theta^{(0)}(z) = 0$ has all the roots outside the unit circle. When the Gauss-Jordan elimination method is applied to invert W, the algorithm with the former starting values needs $\{(s+1)/(q+1)\}^3$ times as many computing operations and $\{(s+1)/(q+1)\}^2$ times as many memory locations as the one with the latter starting values. In particular, if any root of $\phi(z) = 0$ is near the unit circle, s should be chosen to be sufficiently large. Numerical examples show that the two types of the initial values yield the same result. Thus, we recommend using the latter instead of the former.

Table 1.1. A Newton-Raphson solution of $y_t + 0.3y_{t-1} = v_t - 0.7v_{t-1} - 0.18v_{t-2}$

n	$\sigma^{(n)}$	$\psi_1^{(n)}$	$\psi_2^{(n)}$	$d_*^{(n)}$
0	1.419797231	−0.3000000000	0.0900000000	
1	1.246031511	−0.6244565473	0.0855910085	3.5E-1
2	1.102810459	−0.8225054120	0.1025466176	1.3E-1
3	1.034721385	−0.9362260934	0.1136362117	6.2E-2
4	1.007119446	−0.9865683560	0.1186570911	2.7E-2
5	1.000443719	−0.9991573065	0.1199157318	6.6E-3
6	1.000001952	−0.9999962920	0.1199996292	4.4E-4
7	1.000000000	−0.9999999999	0.1200000000	2.0E-6
8	1.000000000	−1.000000000	0.1200000000	3.8E-11
9	1.000000000	−1.000000000	0.1200000000	1.5E-20

Table 1.2. A Newton-Raphson solution of $y_t + 0.3y_{t-1} = v_t - 0.7v_{t-1} - 0.18v_{t-2}$

n	$\sigma^{(n)}$	$\psi_1^{(n)}$	$\psi_2^{(n)}$	$d_*^{(n)}$
0	1.000000000	−1.000000000	0.1200000000	
1	1.000000000	−1.000000000	0.1200000000	8.2E-11
2	1.000000000	−1.000000000	0.1200000000	1.5E-22

The following numerical examples show how fast the algorithm converges. Consider the following ARMA$(1, 2)$ model:

$$y_t + 0.3y_{t-1} = v_t - 0.7v_{t-1} - 0.18v_{t-2}, \quad \sigma^2 = 1.0.$$

Then, $\psi_0 = 1.0$, $\psi_1 = -1.0$, and $\psi_2 = 0.12$. The autocovariances are

$$\begin{aligned} \sigma(0) &= 2.0158241758, & \sigma(1) &= -1.1247472527, \\ \sigma(2) &= 0.15742417582, & \sigma(3) &= -0.047227252747. \end{aligned}$$

The computing result of Algorithm 1.5 with the latter starting values is given in Table 1.1, where

$$d_*^{(n)} = \max_j \mid c_j^{(n)} - c_j^{(n-1)} \mid .$$

If Algorithm 1.5 is applied to an MA(19) model as an approximation to the ARMA$(1, 2)$ process, then $\sigma^{(7)} = 1.000000000$, $\psi_1^{(7)} = -1.000000000$, $\psi_2^{(7)} = 0.1200000000$, and $d_7^{(*)} = 2.0 \times 10^{-6}$. When we use them as starting values for applying Algorithm 1.5 to the ARMA$(1, 2)$ model, we obtain the result in Table 1.2.

The second example illustrates the case that a root of $\phi(z) = 0$ is near the unit circle. Consider the ARMA$(1, 2)$ model

Table 1.3. A Newton-Raphson solution of $y_t + 0.95y_{t-1} = v_t - 0.2v_{t-1} - 0.15v_{t-2}$

n	$\sigma^{(n)}$	$\psi_1^{(n)}$	$\psi_2^{(n)}$	$d_*^{(n)}$
5	1.000001879	−1.149997099	0.9424981743	9.2E-4
6	1.000000000	−1.150000000	0.9425000000	1.9E-6
7	1.000000000	−1.150000000	0.9425000000	1.9E-12

Table 1.4. A Newton-Raphson solution of $y_t + 0.95y_{t-1} = v_t - 0.2v_{t-1} - 0.15v_{t-2}$

n	$\sigma^{(n)}$	$\psi_1^{(n)}$	$\psi_2^{(n)}$	$d_*^{(n)}$
5	1.000001038	−1.149856810	0.9424666852	8.1E-4
6	1.000000000	−1.149999710	0.9424999626	1.0E-6
7	1.000000000	−1.150000000	0.9425000000	2.0E-12
8	1.000000000	−1.150000000	0.9425000000	7.5E-24

$$y_t + 0.95y_{t-1} = v_t - 0.2v_{t-1} - 0.15v_{t-2}, \quad \sigma^2 = 1.0.$$

Then, $\psi_0 = 1.0$, $\psi_1 = -1.15$, and $\psi_2 = 0.9425$. The autocovariances are

$$\sigma(0) = 11.43333333, \quad \sigma(1) = -10.889166667,$$
$$\sigma(2) = 10.194708333, \quad \sigma(3) = -9.6849729167.$$

With the latter choice of starting values, Algorithm 1.5 leads to the result in Table 1.3.

To approximate this ARMA$(1,2)$ process by an MA(s) model, the order should be large because of the near nonstationarity. Even when $s = 19$, $d_*^{(n)}$ does not decrease but oscillates. After 16 iterations, $\sigma^{(16)} = 1.318407178$, $\psi_1^{(16)} = -0.0025145178$, $\psi_2^{(16)} = 0.4897638886$, and $d_*^{(16)} = 7.9$. If we use them as the starting values of Algorithm 1.5, we obtain the result in Table 1.4.

1.4 Estimation

To determine the orders of an ARMA process, it is necessary to obtain preliminary estimates of the parameters for all the possible pairs of orders. Among numerous estimation methods, the method of moments, some least squares methods, and various maximum likelihood methods have been frequently used for the purpose of model identification. We will discuss them in this section. Some iterative least squares estimates will be discussed in Chapter 4.

1.4.1 Extended Yule-Walker Estimates

Let $\{y_1, \ldots, y_T\}$ be a T-realization of a stationary process. Then, we may estimate the ACVF by

$$\hat{\sigma}(k) = \hat{\sigma}(-k) = \frac{1}{T}\sum_{t=1}^{T-k}(y_t - \bar{y})(y_{t+k} - \bar{y}), \quad k = 0, 1, \ldots,$$

where $\bar{y} = \frac{1}{T}\sum_{t=1}^{T} y_t$. It is called the sample ACVF. Jenkins and Watts (1968, Section 5.3) have discussed why it is preferred to any other estimate of the ACVF. The corresponding sample ACRF is defined by

$$\hat{\rho}_k = \frac{\hat{\sigma}(k)}{\hat{\sigma}(0)}, \quad k = 0, \pm 1, \pm 2, \ldots.$$

The asymptotic properties of the sample ACVF and the sample ACRF have been well-known. In order to derive the asymptotic distribution of the sample ACVF we need an additional assumption that the white noise process has the finite fourth-order moment. We define the fourth-cumulant κ_4 by

$$\kappa_4 = Ev_t^4 - 3\sigma^4.$$

Theorem 1.1. *The Asymptotic Distribution of the Sample ACVF*
Let $\{y_t\}$ be a process of the form

$$y_t = \mu + \sum_{j=-\infty}^{\infty} \psi_j v_{t-j},$$

where μ is a constant and $\{v_t\}$ is a sequence of independent and identically distributed random variables with means 0, finite variances σ^2, and finite fourth-cumulant κ_4. If $\sum_{j=-\infty}^{\infty} \mid \psi_j \mid < \infty$, then

$$\sqrt{T}\left(\hat{\sigma}(0) - \sigma(0)\right), \ldots, \sqrt{T}\left(\hat{\sigma}(m) - \sigma(m)\right)$$

are asymptotically normally distributed with means 0 and covariances

$$\lim_{T\to\infty} \operatorname{cov}\left\{\sqrt{T}\left(\hat{\sigma}(r) - \sigma(r)\right), \sqrt{T}\left(\hat{\sigma}(s) - \sigma(s)\right)\right\}$$

$$= \sum_{j=-\infty}^{\infty} \{\sigma(j+r)\sigma(j+s) + \sigma(j-r)\sigma(j+s)\} + \frac{\kappa_4}{\sigma^4}\sigma(r)\sigma(s)$$

$$= 4\pi \int_{-\pi}^{\pi} \cos(\lambda r)\cos(\lambda s) S^2(\lambda) d\lambda + \frac{\kappa_4}{\sigma^4}\sigma(r)\sigma(s),$$

where $S(\lambda)$ is the spectral density of $\{y_t\}$. □

Porat (1987) has shown that the sample ACVF of an ARMA(p, q) process is not always asymptotically efficient. An unbiased estimate of $\sigma(k)$,

$$\frac{T}{T-k}\hat{\sigma}(k),$$

has the asymptotic Cramer-Rao lower bound *if and only if* $0 \leq k \leq p - q$. Thus, if $p < q$, then all the sample autocovariances are asymptotically inefficient. We may obtain an efficient estimate of the ACVF by replacing $\phi_1, \ldots, \phi_p$, $\theta_1, \ldots, \theta_q$, and σ^2 with their maximum likelihood estimates in the EYW equations and Equation (1.2) and solving them.

In contrast to the asymptotic distribution of the sample ACVF, we can derive that of the sample ACRF without the finite fourth-order moment assumption. The following is a generalization of the famous Bartlett's formula (1946) about the asymptotic variance of the sample ACRF.

Theorem 1.2. *The Asymptotic Distribution of the Sample ACRF*
Let $\{y_t\}$ be the process of the form

$$y_t = \mu + \sum_{j=-\infty}^{\infty} \psi_j v_{t-j},$$

where μ is a constant and $\{v_t\}$ is a sequence of independent and identically distributed random variables with means 0 and finite variances σ^2. If

$$\sum_{j=-\infty}^{\infty} | \psi_j | < \infty, \quad \sum_{j=-\infty}^{\infty} | j | \psi_j^2 < \infty,$$

then

$$\sqrt{T}(\hat{\rho}_1 - \rho_1), \ldots, \sqrt{T}(\hat{\rho}_m - \rho_m)$$

are asymptotically normally distributed with means 0 and covariances

$$\lim_{T\to\infty} \operatorname{cov}\left\{\sqrt{T}(\hat{\rho}_r - \rho_r), \sqrt{T}(\hat{\rho}_s - \rho_s)\right\}$$
$$= \sum_{j=-\infty}^{\infty} \{\rho_{j+r}\rho_{j+s} + \rho_{j-r}\rho_{j+s} + 2\rho_r\rho_s\rho_j^2$$
$$-2\rho_r\rho_j\rho_{j+s} - 2\rho_s\rho_j\rho_{j+r}\}. \quad \square$$

For more details about the asymptotic properties of the sample ACVF and the sample ACRF including derivations of Theorems 1.1 and 1.2, readers may consult the references in Section 1.6.

If the observations are from the ARMA(p, q) model (1.1), we can estimate the parameters $\phi_1, \ldots, \phi_p$, $\theta_1, \ldots, \theta_q$, and σ^2 through the method of moments, i.e., we substitute the sample ACVF for the ACVF in the EYW equations and Equation (1.2), and solve them using the algorithms presented in Section 1.3. Throughout this book they will be called the EYW estimates and will be denoted by $\hat{\phi}_1, \ldots, \hat{\phi}_p$, $\hat{\theta}_1, \ldots, \hat{\theta}_q$, and $\hat{\sigma}^2$. The EYW estimates of $\phi_{k,1}^{(i)}, \ldots, \phi_{k,k}^{(i)}$ are defined as the solutions of the simultaneous equations

$$\hat{\rho}_j = \hat{\phi}_{k,1}^{(i)} \hat{\rho}_{j-1} + \cdots + \hat{\phi}_{k,k}^{(i)} \hat{\rho}_{j-k}, \quad j = i+1, \ldots, i+k.$$

Denote the EYW estimates of $\boldsymbol{\phi}(k,i)$ by $\hat{\boldsymbol{\phi}}(k,i)$. The nonsingularity of $\Sigma(k,i)$ and the consistency of the sample ACVF imply the consistency of $\hat{\boldsymbol{\phi}}(k,i)$.

Theorem 1.3. *The Consistency of the EYW Estimates*
Let $\{y_1, \ldots, y_T\}$ be a T-realization from the ARMA(p,q) model (1.1). Then $\hat{\boldsymbol{\phi}}(k,i)$ is consistent to $\boldsymbol{\phi}(k,i)$ for $(k,i) \in \{(k,q) \mid k = p, p+1, \ldots\} \bigcup \{(p,i) \mid i = q, q+1, \ldots\}$. □

If $k = p+1, p+2, \ldots$, $i = q+1, q+2, \ldots$, then $\hat{\boldsymbol{\phi}}(k,i)$ is inconsistent because $\Sigma(k,i)$ is singular.

Let $\{y_1, \ldots, y_T\}$ be a T-realization from an AR(p) model. Then, $\hat{\phi}_{k,1}, \ldots, \hat{\phi}_{k,k}$ and $\hat{\sigma}_k^2$ are called the YW estimates. Mann and Wald (1943) have derived the asymptotic distribution of $\hat{\phi}_{p,1}, \ldots, \hat{\phi}_{p,p}$. The following is a generalization.

Theorem 1.4. *The Asymptotic Distribution of the YW Estimates*
Let $\{y_t\}$ be from the AR(p) model (1.1) with $q = 0$. If $m \geq p$, then

$$T^{1/2}(\hat{\phi}_{m,1} - \phi_{m,1}), \ldots, T^{1/2}(\hat{\phi}_{m,m} - \phi_{m,m})$$

are asymptotically normally distributed with means 0 and covariance matrix $\sigma^2 \Sigma^{-1}(m,0)$. □

If $q = 0$, then $\hat{\sigma}(0), \ldots, \hat{\sigma}(p)$ are asymptotically efficient. Thus, the YW estimates are asymptotically efficient. But, for small T, the YW estimates are biased toward 0 due to bad end effects. It will be discussed in the next subsection.

Choi (1990d) has derived the asymptotic distribution of the EYW estimates, which is a generalization of Theorem 1.4.

Theorem 1.5. *The Asymptotic Distribution of the EYW Estimates*
Let $\{y_1, \ldots, y_T\}$ be a T-realization from the ARMA(p,q) model (1.1). If $k = p, p+1, \ldots$, then $\sqrt{T}\{\hat{\boldsymbol{\phi}}(k,q) - \boldsymbol{\phi}(k,q)\}$ is asymptotically normally distributed with mean $\mathbf{0}$ and covariance matrix

$$\sigma^2 \Sigma(k,q)^{-1} \Xi \Sigma^{-1}(k,q)^t,$$

where the (r, s) element of Ξ is $\boldsymbol{\theta}_*^t \Sigma(q+1, r-s)\boldsymbol{\theta}_*$. □

1.4.2 Maximum Likelihood Estimates

Consider the ARMA(p, q) process (1.1) with an additional assumption that the white noise process $\{v_t\}$ is Gaussian. Let $\mathbf{y} = (y_1, \ldots, y_T)^t$ be an observation vector. Because it is challenging to obtain the exact likelihood function of $\mathbf{y}$, some variant maximum likelihood (ML) estimates of $\boldsymbol{\phi}$, $\boldsymbol{\theta}$, and σ^2 have been suggested.

If $\mathbf{y}_* = (y_{1-p}, y_{2-p}, \ldots, y_0)^t$ and $\mathbf{v}_* = (v_{1-q}, v_{2-q}, \ldots, v_0)^t$ are given, then the conditional log-likelihood function is

$$l_*(\boldsymbol{\phi}, \boldsymbol{\theta}, \sigma^2) = -\frac{T}{2}\ln(2\pi\sigma^2) - \frac{S_*(\boldsymbol{\phi}, \boldsymbol{\theta})}{2\sigma^2},$$

where

$$S_*(\boldsymbol{\phi}, \boldsymbol{\theta}) = \sum_{t=1}^{T} v_t^2(\boldsymbol{\phi}, \boldsymbol{\theta} \mid \mathbf{y}_*, \mathbf{v}_*, \mathbf{y}).$$

The quantities maximizing $l_*(\boldsymbol{\phi}, \boldsymbol{\theta}, \sigma^2)$ are called the *conditional ML estimates.* Some suggestions have been made for specifying the initial vectors $\mathbf{y}_*$ and $\mathbf{v}_*$. If we let $\mathbf{y}_* = \mathbf{0}$ and $\mathbf{v}_* = \mathbf{0}$, then the conditional ML estimates of $\boldsymbol{\phi}$ and $\boldsymbol{\theta}$ are ordinary least squares (OLS) estimates.

If we consider only AR models, then the normal equations of the least squares problem are the YW equations. Thus, if we obtain the OLS estimates by solving the normal equations, then they are the YW estimates. However, numerical analysts have preferred minimizing the sum of squares directly to solving the normal equations. One of the reasons is that the condition number of the normal equations is intrinsically the square of that of the existent OLS problem. For more details, readers may refer to Golub and Van Loan (1989, Chapter 5). In this book the OLS estimates mean to minimize the sum of squares directly, not to solve the normal equations. There is an efficient orthogonalization method to minimize $S_*(\boldsymbol{\phi}, \boldsymbol{\theta})$ directly. It is a lattice method and its numerical stability has been shown by Cybenko (1984). There are more references about lattice algorithms and Levinson-Durbin type algorithms in Section 1.6.

Burg (1975) has proposed an algorithm to estimate the parameters of the AR model using the observations not via the sample ACVF. Define the forward and the backward innovations by

$$v_{k,t} = y_t - \sum_{j=1}^{k} \phi_{k,j} y_{t-j}, \quad t = k+1, \ldots, T,$$

$$u_{k,t} = y_t - \sum_{j=1}^{k} \phi_{k,j} y_{t+j}, \quad t = 1, \ldots, T-k,$$

respectively. We can calculate them by the recursive formulas

$$v_{k,t} = v_{k-1,t} + \phi_{k,k} u_{k-1,t-k}, \quad t = k+1, \ldots, T,$$

$$u_{k,t} = u_{k-1,t} + \phi_{k,k} v_{k-1,t+k}, \quad t = 1, \ldots, T-k.$$

Burg's estimate $\check{\phi}_{k,k}$ is defined by the one minimizing

$$\sum_{t=k+1}^{T} v_{k,t}^2 + \sum_{t=1}^{T-k} u_{k,t}^2.$$

It becomes

$$\check{\phi}_{k,k} = \frac{2 \sum_{t=k+1}^{T} v_{k-1,t} u_{k-1,t-k}}{\sum_{t=k+1}^{T} v_{k-1,t}^2 + \sum_{t=1}^{T-k} u_{k-1,t}^2}.$$

Burg's estimates of the other AR parameters are

$$\check{\phi}_{k,j} = \check{\phi}_{k-1,j} - \check{\phi}_{k,k} \check{\phi}_{k-1,k-j}, \quad j = 1, \ldots, k-1.$$

Readers may refer to Andersen (1974, 1978), Morf, Vieira, and Kailath (1978), and Arcese (1983) for derivations of Burg's algorithm.

The YW estimates, the OLS estimates, and Burg's estimates are asymptotically equivalent. However, there are some differences among the three estimates in small sample cases. The YW estimates are sometimes out of the stationary region unlike Burg's. In his letter to the present author, E. J. Hannan has mentioned that if the YW estimates are outside of the stationarity region, it must due to numerical instability. It has been shown by Tjøstheim and Paulsen (1983) and Paulsen and Tjøstheim (1985) that the YW estimates are more biased than the OLS estimates and Burg's estimates for small T. They have concluded that the YW estimates are inferior to the OLS estimates and the Burg-type estimates. Lysne and Tjøstheim (1987) have mentioned that Burg's estimates behave slightly better than the OLS estimates based on their simulation results on the low order AR spectral density estimates. In contrast, Marple (1980) has reported that Burg's estimates have a tendency to produce artificial splitting of spectral peaks. As shown by Cybenko (1980, 1983, 1984), Burg's estimates and the OLS estimates are more desirable with regard to numerical stability, particularly if a root of $\phi(z) = 0$ is near the unit circle. There are more references about biases of the estimates in Section 1.6.

The stationarity assumption implies that $\phi(B)y_t = \theta(B)v_t$ has the same autocovariance structure as

$$\phi(B^{-1})y_t = \theta(B^{-1})a_t,$$

where $\{a_t\}$ is a white noise process. Box and Jenkins (1976, p. 197) referred to it as the backward form of the process or, in short, the backward process. They proposed to estimate the starting values $y_0, y_{-1}, \ldots,$ using the backward process, and then to maximize the unconditional log-likelihood function

$$l(\boldsymbol{\phi}, \boldsymbol{\theta}, \sigma^2) = -\frac{T}{2}\ln(2\pi\sigma^2) - \frac{S(\boldsymbol{\phi}, \boldsymbol{\theta})}{2\sigma^2},$$

where

$$S(\boldsymbol{\phi}, \boldsymbol{\theta}) = \sum_{t=-\infty}^{T} v_t^2(\boldsymbol{\phi}, \boldsymbol{\theta} \mid \mathbf{y}).$$

The quantities maximizing $l(\boldsymbol{\phi}, \boldsymbol{\theta}, \sigma^2)$ are called the *unconditional ML estimates.* Because the estimates of $\phi_1, \ldots, \phi_p$ and $\theta_1, \ldots, \theta_q$ are equivalent to minimizing $S(\boldsymbol{\phi}, \boldsymbol{\theta})$, they are also the *unconditional least squares estimates.*

It is an intricate problem to obtain the ML estimates of ARMA processes. Many time series analysts have attempted to solve the problem and there have been some progress. Interested readers may consult the references in Section 1.6. It is known that the ML estimates are consistent and asymptotically normally distributed. In this book any type of the ML estimates will be denoted by $\tilde{\phi}_1, \ldots, \tilde{\phi}_p, \tilde{\theta}_1, \ldots, \tilde{\theta}_q$, and $\tilde{\sigma}^2$.

Because it is beyond the scope of this book to discuss other ARMA estimation methods, we just present some references in Section 1.6.

1.5 Nonstationary Processes

We can categorize ARMA processes according to the location of the roots of $\phi(z) = 0$. In stochastic processes analysis a process is called strictly stationary *if and only if* its joint distributions remain the same as time passes. (See, e.g., Doob [1953, p. 94].) If all the roots of $\phi(z) = 0$ are outside the unit circle, then y_t can be written as an infinite linear combination of $v_t, v_{t-1}, \ldots$, i.e., y_t is independent of $v_{t+1}, v_{t+2}, \ldots$, and the process is strictly stationary. Otherwise, y_t may depend on the future innovations $v_{t+1}, v_{t+2}, \ldots$. For example, if all the roots of $\phi(z) = 0$ are inside the unit circle, then y_t can be represented by an infinite linear combination of $v_{t-q+1}, v_{t-q+2}, \ldots$, i.e., y_t is dependent of future innovations $v_{t+1}, v_{t+2}, \ldots$, but the process is strictly stationary. Moreover, if $\{y_t\}$ is strictly stationary and at least one root of $\phi(z) = 0$ is 1, then all the values of the process are the same with probability 1. (See, e.g., T. W. Anderson [1971, p. 172].) It is natural to assume that the current observation y_t is independent of the future innovations $v_{t+1}, v_{t+2}, \ldots$. It is Box and Jenkins' definition of the stationarity (1976), i.e., an ARMA process is stationary *if and only if* all the roots of $\phi(z) = 0$ are outside the unit circle. In this book we follow Box and Jenkins' definition of the stationarity.

If we assume that $y_t = 0$ and $v_t = 0$ for $t = 0, -1, -2, \ldots$, then we can write y_t as

$$y_t = \sum_{j=0}^{t-1} \psi_j v_{t-j},$$

where ψ_j is the jth coefficient of B^j in

$$\psi(B) = \theta(B)\phi^{-1}(B) = \sum_{j=0}^{\infty} \psi_j B^j.$$

More specifically,

$$\begin{aligned}
\psi_0 &= 1, \\
\psi_1 &= \psi_0\phi_1 - \theta_1, \\
\psi_2 &= \psi_0\phi_2 + \psi_1\phi_1 - \theta_2, \\
&\vdots \\
\psi_q &= \psi_0\phi_q + \cdots + \psi_{q-1}\phi_1 - \theta_q, \\
\psi_j &= \psi_0\phi_j + \cdots + \psi_{j-1}\phi_1 - \theta_j, \\
&\qquad j = q+1, q+2, \ldots,
\end{aligned}$$

where $\phi_{p+1} = \phi_{p+2} = \cdots = 0$. If all the roots of $\phi(z) = 0$ are outside the unit circle, i.e., the ARMA process is stationary, then $\sum_{j=0}^{\infty} \psi_j$ converges absolutely; otherwise, it diverges. Particularly, if any root of $\phi(z) = 0$ is inside the unit circle, then $\mid \psi_j \mid$ increases exponentially as j increases. Such a process is called explosive. Huzii (1981) has presented a method to test whether an AR model is explosive or not. For the case of an AR(1) model, he has defined

$$\pi_h = \frac{E y_t^2 y_{t+h}}{E y_t^2 y_{t-h}}, \quad h > 0,$$

and has shown that

$$\begin{aligned}
\mid \pi_h \mid &> 1 \text{ if } \mid \phi_1 \mid < 1, \\
\mid \pi_h \mid &< 1 \text{ if } \mid \phi_1 \mid > 1.
\end{aligned}$$

He has also suggested the use of a consistent estimate of π_h defined by

$$\hat{\pi}_h = \frac{\sum_{t=1}^{T-h} y_t^2 y_{t+h}}{\sum_{t=h+1}^{T} y_t^2 y_{t-h}}$$

as a test statistic.

Recently some econometricians have paid attention to utilizing ARMA processes with unit roots of $\phi(z) = 0$. Such processes have appeared in stock market and commodity market data as well as aggregated time series. Phillips and Perron (1988) provided a brief history of the use of the unit root models in economics. If all the roots of $\phi(z) = 0$ are on or outside the unit circle and if some of them are on the unit circle, then we will call the ARMA process nonstationary in this book. It is also called unstable. During the last two decades there have been some progresses in analyzing unstable ARMA processes. Especially, the three papers by Tiao and Tsay (1983a) and Tsay and Tiao (1984, 1985), hereafter referred to as TT-83, TT-84, and TT-85, respectively, have constituted a comprehensive study of nonstationary ARMA processes. Most of the results in this section are from them. There are more references about nonstationary ARMA processes in Section 1.6.

1.5.1 Sample ACRF of a Nonstationary Process

Hereafter, when we consider a nonstationary ARMA(p, q) model

$$\phi(B)y_t = \theta(B)v_t, \tag{1.3}$$

we assume that all the roots of $\phi(z) = 0$ are on or outside the unit circle and $\{v_t\}$ is a sequence of independent and identically distributed random variables satisfying

$$E(v_t) = 0, \quad E(v_t^2) = \sigma^2 \ (< \infty), \quad E(v_t^4) = 3\sigma^4 + \kappa_4 \ (< \infty).$$

The finite fourth-cumulant assumption does not play an important role in analyzing real time series, but it is necessary to derive some theorems of the nonstationary process. Also, we assume that $y_t = 0$ and $v_t = 0$ for $t = 0, -1, -2, \ldots$, and the coefficient ψ_j is the same as the one defined before.

Property 1.1. Let $\{y_t\}$ be from the nonstationary ARMA(p, q) model (1.3). Then

$$E\left(\sum_{t=1}^{T} y_t^2\right) = \sigma^2 \sum_{t=1}^{T} (T - t + 1)\, \psi_{t-1}^2,$$

$$\begin{aligned}\operatorname{var}\left(\sum_{t=1}^{T} y_t^2\right) &= (2\sigma^2 + \kappa_4) \sum_{t=1}^{T} \left(\sum_{l=0}^{t-1} \psi_l^2\right)^2 \\ &\quad + 4\sigma^4 \sum_{t=1}^{T-1} \sum_{j=t+1}^{T} \left(\sum_{l=0}^{t-1} \psi_l \psi_{l+j-t}\right)^2 . \qquad \square\end{aligned}$$

If the process is stationary, Property 1.1 and the Cesáro summation imply

$$\lim_{T\to\infty} E\left(\frac{1}{T} \sum_{t=1}^{T} y_t^2\right) = \sigma^2 \sum_{l=0}^{\infty} \psi_l^2.$$

If the process is nonstationary, then $\sum_{l=0}^{\infty} \psi_l^2$ diverges and the ACVF cannot be defined. However, we can extend the sample ACRF to the nonstationary case. Let $\alpha_1, \ldots, \alpha_I$ be the distinct roots of $\phi(z) = 0$ having the highest multiplicity m among the roots on the unit circle. Denote the other roots on the unit circle by $\beta_1, \ldots, \beta_J$, and let

$$U_\alpha(B) = \prod_{i=1}^{I} \left(1 - \frac{1}{\alpha_i} B\right),$$

$$U_\beta(B) = \prod_{j=1}^{J} \left(1 - \frac{1}{\beta_j} B\right).$$

Then, the polynomial corresponding to all the roots on the unit circle is

$$U(B) = U_\alpha^m(B) U_\beta(B).$$

If the order of $U(B)$ is denoted by d, then

$$d = mI + J.$$

Property 1.2. Let $\{y_t\}$ be from the nonstationary ARMA(p, q) model (1.3). If $m > 0$, then

$$\sum_{t=1}^{T} y_t^2 = O_p(T^{2m}),$$

$$\frac{1}{\sum_{t=1}^{T} y_t^2} = O_p\left(\frac{1}{T^{2m}}\right). \quad \square$$

Based on Property 1.2, we define the sample autocorrelations of the nonstationary process $\{y_t\}$ as

$$\hat{\rho}_k = \frac{\sum_{t=1}^{T} y_t y_{t+k}}{\sum_{t=1}^{T} y_t^2}, \quad k = 0, 1, \ldots.$$

Clearly, it is a generalization of the sample ACRF of a stationary process to a nonstationary process. It has been shown (Quinn [1980a], Findley [1980], TT-83) that the asymptotic behaviors of the sample ACRF of the nonstationary process are dominated by the nonstationary root with the highest multiplicity as follows.

Property 1.3. Let $\{y_t\}$ be from the nonstationary ARMA(p, q) model (1.3). If $m > 0$, then

$$U_\alpha(B) \hat{\rho}_k = 0. \quad \square$$

1.5.2 Iterated Least Squares Estimates

As discussed before, one of the most popular estimation methods in time series analysis is the OLS approach. It has been shown by Mann and Wald (1943) and TT-83 that the OLS estimates are consistent to the AR parameters if the underlying process is either from a stationary AR(p) process or from a purely nonstationary ARMA process, i.e., all the roots are on the unit circle. Otherwise, the OLS estimates are inconsistent. For a further explanation, we fit an AR(l) model

$$y_t = \beta_1 y_{t-1} + \cdots + \beta_l y_{t-l} + u_t, \quad t = l+1, \ldots, T,$$

where $\{u_t\}$ is assumed to be a white noise process. The OLS estimate of $\boldsymbol{\beta} = (\beta_1, \ldots, \beta_l)^t$ is

$$\hat{\boldsymbol{\beta}} = H_{T,l}^{-1} \sum_{t=l+1}^{T} \boldsymbol{w}_{t-1,l} y_t,$$

where

$$\boldsymbol{w}_{t,l} = (y_t, y_{t-1}, \ldots, y_{t-l+1})^t,$$

$$H_{T,l} = \sum_{t=l+1}^{T} \boldsymbol{w}_{t-1,l} \boldsymbol{w}_{t-1,l}^t.$$

The existence of the OLS estimate depends on the nonsingularity of $H_{T,l}$.

Property 1.4. Let $\{y_t\}$ be from the ARMA(p, q) model (1.3). For a positive integer l, if $T \geq 2l$, then $H_{T,l}$ is positive definite with probability 1. □

Using Properties 1.1-1.4, TT-83 have shown the inconsistency of the OLS estimates of mixed ARMA processes.

Property 1.5. Let $\{y_t\}$ be a nonstationary ARMA(p, q) process with $d = mI + J > 0$. For $l = 0, 1, \ldots,$ the OLS estimates of the AR($d + l$) regression are inconsistent, unless $l \geq p - d$ and $q = 0$. □

TT-84 proposed a consistent iterative least squares procedure of the AR parameters when the OLS estimates are inconsistent. Let $\{y_1, \ldots, y_T\}$ be a T-realization from the ARMA(p, q) model. First, we fit a pure AR(p) model

$$y_t = \sum_{l=1}^{p} \phi_{l(p)}^{(0)} y_{t-l} + e_{p,t}^{(0)}, \quad t = p+1, \ldots, T.$$

Here the superscript (0) means a pure autoregression, the subscript (p) denotes the order of AR fitting, and $e_{p,t}^{(0)}$ is the corresponding innovation term. As discussed above, $\hat{\phi}_{l(p)}^{(0)}$ is consistent to ϕ_l, only if the underlying process is either an AR(p) process or a purely nonstationary ARMA process. As the second step, we consider the case that $\hat{\phi}^{(0)}_{l(p)}$ is inconsistent. Because the estimated residuals $\{\hat{e}^{(0)}_{p,t}\}$ are not white noise, they contain some information about the process $\{y_t\}$. This makes it possible to define the first iterated AR(p) regression as

$$y_t = \sum_{l=1}^{p} \phi^{(1)}_{l(p)} y_{t-l} - \theta^{(1)}_{1(p)} \hat{e}^{(0)}_{p,t-1} + e^{(1)}_{p,t}, \quad t = p+2, \ldots, T,$$

where the superscript (1) means the first iterated autoregression and $e^{(1)}_{p,t}$ denotes the corresponding innovation term. It is shown that the least squares estimates $\hat{\phi}^{(1)}_{1(p)}, \ldots, \hat{\phi}^{(1)}_{p(p)}$ are consistent, either if $q = 0, 1$ or if all the zeros of $\phi(z)$ are on the unit circle. Through the similar procedure we can construct the iterated least squares (ILS) estimates of the AR parameters for $k = 1, 2, \ldots$ and $i = 0, 1, \ldots$. The ith iterated AR(k) regression is defined as

$$y_t = \sum_{l=1}^{k} \phi^{(i)}_{l(k)} y_{t-l} - \sum_{j=1}^{i} \theta^{(i)}_{j(k)} \hat{e}^{(i-j)}_{k,t-j} + e^{(i)}_{k,t}, \quad t = k+i+1, \ldots, T,$$

where

$$\hat{e}^{(i)}_{k,t} = y_t - \sum_{l=1}^{k} \hat{\phi}^{(i)}_{l(k)} y_{t-l} + \sum_{j=1}^{i} \hat{\theta}^{(i)}_{j(k)} \hat{e}^{(i-j)}_{k,t-j}$$

is the estimated residual of the ith iterated AR(k) regression and $\phi^{(i)}_{1(k)}$, $\ldots$, $\phi^{(i)}_{k(k)}$ are the corresponding ILS estimates.

TT-84 presented a recursive algorithm for the ILS estimates.

Algorithm 1.6 For $k = 1, 2, \ldots$ and $i = 1, 2, \ldots$,

$$\hat{\phi}^{(i)}_{j(k)} = \hat{\phi}^{(i-1)}_{j(k+1)} - \hat{\phi}^{(i-1)}_{j-1(k)} \frac{\hat{\phi}^{(i-1)}_{k+1(k+1)}}{\hat{\phi}^{(i-1)}_{k(k)}}, \quad j = 1, \ldots, k,$$

where $\hat{\phi}^{(i)}_{0(k)} = -1$. □

It equals Algorithm 1.4 for the EYW estimates of the AR parameters up to initial values. Using Algorithm 1.6, TT-84 showed the consistency of the ILS estimates of nonstationary ARMA processes.

Theorem 1.6. *The Consistency of the ILS Estimates*
Let $\{y_1, \ldots, y_T\}$ be a T-realization from the ARMA(p, q) model (1.3). Assume that $\phi_j = 0$ for $j = p+1, p+2, \ldots$. If $(k, i) \in \{(k, q) \mid k =$

$p, p+1, \ldots\} \bigcup \{(p,i) \mid i = q, q+1, \ldots\}$, then

$$\hat{\phi}^{(i)}_{j(k)} = \phi_j + O_p\left(\frac{1}{\sqrt{T}}\right), \quad j = 1, \ldots, k.$$

The term $O_p(1/\sqrt{T})$ becomes $O_p(1/T)$ if the process is purely nonstationary and $k = p$. □

From Algorithms 1.4 and 1.6 we anticipate that the EYW estimates and the ILS estimates would be asymptotically equivalent as long as the underlying ARMA process is stationary.

Theorem 1.7. Let $\{y_1, \ldots, y_T\}$ be a T-realization from the stationary ARMA(p,q) model (1.1). Then, for $j = i+1, \ldots, i+k$,

$$\hat{\rho}_j - \sum_{l=1}^{k} \hat{\phi}^{(i)}_{l(k)} \hat{\rho}_{j-l} = O_p\left(\frac{1}{T}\right). \quad \square$$

Using Cramer's rule we can derive the following.

Corollary 1.1. Let $\{y_1, \ldots, y_T\}$ be a T-realization from the stationary ARMA(p,q) model (1.1). If $(k,i) \in \{(k,q) \mid k = p, p+1, \ldots\} \bigcup \{(p,i) \mid i = q, q+1, \ldots\}$, then

$$\hat{\phi}^{(i)}_{j(k)} = \hat{\phi}^{(i)}_{k,j} + O_p\left(\frac{1}{T}\right), \quad j = 1, \ldots, k. \quad \square$$

Corollary 1.1 means the EYW estimates of the AR parameters are asymptotically equivalent to the ILS estimates if the underlying process is stationary. However, the asymptotic equivalence in Corollary 1.1 breaks down for nonstationary ARMA processes, especially either when some roots of $\phi(z) = 0$ are on the unit circle and some are outside it or when all the roots are on the unit circle but not all the roots are distinct.

Finally, it should be noted that most of the identification methods for stationary ARMA processes, which are discussed in this book, can be applied to nonstationary processes if the ILS estimates are used instead of the EYW estimates and the ML estimates.

1.6 Additional References

- (Section 1.1) For the stationarity and the invertibility conditions, refer to Wise (1956), Barndorff-Nielsen and Schou (1973), Pagano (1973, 1974), Ramsey (1974), O. D. Anderson (1975b, 1977b, 1978),

Granger and Andersen (1978), Hallin (1980, 1981), Piccolo (1982), Monahan (1984), Findley (1986), Hallin, Lefevre, and Puri (1988), and the references therein. Particularly, Piccolo (1982) presented the hypervolume of the stationarity and the invertibility regions for the ARMA model and commented on the severity of the constraints for higher order models.

- (Section 1.1) For non-Gaussian ARMA processes, refer to Granger (1979), Davies, Spedding, and Watson (1980), Lawrance and Lewis (1980), Pierce (1985), Martin and Yohai (1985), Findley (1986), Hallin, Lefevre and Puri (1988), Priestley (1988), Harvey (1989, pp. 348-362), Damsleth and El-Shaarawi (1989), Diggle and Zeger (1989), and the references therein.

- (Section 1.1) For varying coefficient ARMA models, refer to Nicholls and Pagan (1985), Hallin and Ingenbleek (1983), Hallin (1984, 1986), and the references therein.

- (Section 1.1) There are some related works to Equation (1.2) by McLeod (1975) and Mittnik (1990).

- (Section 1.2) For the history of using the ARMA model, refer to Wold (1938, 1966), Rudra (1954), Brillinger (1981, pp. 9-13), and Lauritzen (1981).

- (Section 1.2) There are more survey papers of recent developments in time series analysis such as Jenkins (1965), Kailath (1974), Makridakis (1976, 1978), O. D. Anderson (1977a), Chatfield (1977), Hopwood and Newbold (1980), Kay and Marple (1981), Cox (1981), Abraham and Ledolter (1986), and Pino, Morettin and Mentz (1987).

- (Section 1.2) There are more books of collected papers in time series analysis edited by Rosenblatt (1963), Harris (1967), Parzen (1967), Mehra and Lainiotis (1976), Zellner (1978), Childers (1978), Findley (1978, 1981), Haykin (1979), Makridakis and Wheelwright (1979), Brillinger and Tiao (1980), O. D. Anderson (1980b, 1980c, 1982a, 1982b, 1983, 1984a, 1984b, 1985a, 1985b), O. D. Anderson and Perryman (1981, 1982), Haykin and Cadzow (1982), Box, Leonard, and Wu (1983), Mandrekar and Salehi (1983), Parzen (1983b), Franke, Härdle, and Martin (1984), Wegman and Smith (1984), Gani and Priestley (1986), Kesler (1986), C. H. Chen (1989), and Bittanti (1989).

- (Section 1.3) For more details about Algorithm 1.2, refer to Bednar and Roberts (1985), Takemura (1984), Franke (1985a), and Choi (1991a). Algorithm 1.2 has been generalized to a vector ARMA process by Whittle (1963), Wiggins and Robinson (1965), Rissanen (1973), Watson (1973), Akaike (1973b), Trench (1974), Jong (1976),

Morf, Vieira, and Kailath (1978), and Choi (1990c). Also, refer to Section 4.2 of this book.

- (Section 1.4) For more details about the asymptotic properties of the sample ACVF and the sample ACRF, readers may refer to some standard textbooks of time series analysis such as T. W. Anderson (1971, Chapter 8), Fuller (1976, Chapter 6), Priestley (1981, Section 5.3), and Brockwell and Davis (1987, Chapter 7). For derivations of their distributions, refer to R. L. Anderson (1942), Anderson and Rubin (1964), Ramasubban (1972), Hannan and Heyde (1972), Hannan (1976), Roy (1989), and the references therein. For uniform convergence of the sample ACVF and the sample ACRF, refer to Hannan (1974), An, Chen, and Hannan (1982), Hannan and Kavalieris (1983b), Hannan and Deistler (1988, Section 5.3), and the references therein. Their results will be useful for studying asymptotic properties of the orders selected by penalty function methods, which will be discussed in Chapter 3.

- (Section 1.4) For the lattice method, refer to Cybenko (1983) and Caines (1988, pp. 194-198). Actually there are a lot of lattice algorithms and Burg's algorithm is one of them. For more details, readers may refer to survey papers by Friedlander (1982a, 1982b). Also, refer to Lee, Morf, and Friedlander (1981), Makhoul (1981), Lee, Friedlander, and Morf (1982), Ensor and Newton (1990), and the special volume of *IEEE Transactions on Acoustics, Speech, and Signal Processing*, **ASSP-29**, No. 3 (1981). Barrodale and Erickson (1980), Marple (1980), Huang (1990b), and Strobach (1990, Chapter 7) studied Levinson-Durbin type algorithms for AR model fitting of stationary and nonstationary processes.

- (Section 1.4) For more details about biases of the YW estimates, Burg's estimates, and the OLS estimates, refer to White (1961), Shenton and Johnson (1965), Tanaka (1984), Yamamoto and Kunitomo (1984), Kunitomo and Yamamoto (1985), Nicholls and Pope (1988), Shaman and Stine (1988), Stine and Shaman (1990), Pope (1990), and the references therein. About convergence of the OLS estimates, refer to Shibata (1977).

- (Section 1.4) For the ML estimates and the related topics, refer to Whittle (1951, 1952a, 1953b, 1962), Walker (1964), Reeves (1972), Tunnicliffe-Wilson (1973), Akaike (1973a), Åström and Söderström (1974), Newbold (1974), T. W. Anderson (1975, 1977), Dunsmuir and Hannan (1976), Box and Jenkins (1976), Nicholls (1976, 1977), Osborn (1976, 1977), Godolphin (1977, 1978, 1980a, 1984), Ali (1977), Cooper and Thompson (1977), Kohn (1977), McLeod (1975), Deistler, Dunsmuir, and Hannan (1978), Phadke and Kedem (1978),

Pham (1978, 1979, 1984a, 1986, 1987), Ansley (1979), McDunnough (1979), Harvey and Phillips (1979), Ljung and Box (1979), Nicholls and Hall (1979), Pearlman (1980), T. W. Anderson and Mentz (1980), Kabaila (1980, 1983), Cooper and Wood (1981), Cryer and Ledolter (1981), Ljung (1982), Godolphin and Gooijer (1982), Ansley and Kohn (1983), Spliid (1983), Godolphin and Unwin (1983), Tanaka (1984, 1986), Kulperger (1985), Wincek and Reinsel (1986), Pötscher (1987), Shea (1987), Cernuschi-Frías and Rogers (1988), Dahlhaus and Pötscher (1989), and the references therein. Recently some methods have been presented to calculate the likelihood function via the Kalman filter, which was proposed by Kalman (1960, 1963) and Kalman and Bucy (1961). For more details, refer to Hannan and Deistler (1988, Chapter 6).

- (Section 1.4) There are some other ARMA estimation methods. Walker (1962) proposed to maximize the likelihood function of the sample ACVF. Box and Jenkins (1976) proposed to estimate the parameters using the method of moments and Wilson's algorithm (1969) together. Konvalinka and Matausek (1979) presented a method based on least squares input-output analysis. Some estimation methods using spectral analysis were proposed by Hannan (1970a, 1979), Parzen (1971), Nicholls (1972, 1973), and T. W. Anderson (1975). For robust estimation methods, refer to Denby and Martin (1979), Martin (1980, 1981), Martin and Yohai (1985), Bustos and Yohai (1986), Masarotto (1987), Li and Hui (1989), and the references therein.

- (Section 1.5) The explosive ARMA model has been studied by Rao (1961), Stigum (1974), Fuller and Hasza (1981), O. D. Anderson (1990) and Huang (1990a, 1990b). For unstable ARMA processes, refer to O. D. Anderson (1975a), Fuller (1976, pp. 366-385; 1985), Roy (1977), Dickey and Fuller (1979, 1981), Hasza and Fuller (1979), Kawashima (1980), Hasza (1980), Fuller and Hasza (1981), Fuller, Hasza, and Goebel (1981), Evans and Savin (1981, 1984), Sargan and Bhargava (1983), Kay (1983), Ahtola and Tiao (1984, 1987a, 1987b), Solo (1984), Yajima (1985), Said and Dickey (1984, 1985), Parzen (1986), Bhargava (1986), Phillips (1987a, 1987b), Stoica and Nehorai (1986, 1987), Chan and Wei (1987, 1988), Sims (1988), Cressie (1988), Chan (1988, 1990), Hall (1989), Perron (1989), Pukkila (1989), Schwert (1989), Porter-Hudak (1990), and the references therein. Tsay and Tiao (1990) considered the asymptotic properties of multivariate nonstationary ARMA processes.

2

The Autocorrelation Methods

2.1 Box and Jenkins' Method

It is known (see, e.g., T. W. Anderson [1971, pp. 463-495]) that, for large T, the bias of the sample ACRF is

$$E(\hat{\rho}_k) - \rho_k = -\frac{1}{T}\sum_{j=-\infty}^{\infty} \rho_j + o\left(\frac{1}{T}\right).$$

Theorem 1.2 shows that the covariance $\operatorname{cov}(\hat{\rho}_r, \hat{\rho}_s)$ of the sample ACRF is

$$\frac{1}{T}\sum_{j=-\infty}^{\infty} \{\rho_{j+r}\rho_{j+s} + \rho_{j-r}\rho_{j+s} + 2\rho_r\rho_s\rho_j^2$$
$$-2\rho_r\rho_j\rho_{j+s} - 2\rho_s\rho_j\rho_{j+r}\} + o\left(\frac{1}{T}\right).$$

Particularly, if the process is Gaussian, the variance and the covariance become

$$\operatorname{var}(\hat{\rho}_k) \simeq \frac{1}{T}\sum_{j=-\infty}^{\infty} \{\rho_j^2 + \rho_{j+k}\rho_{j-k}\},$$

$$\operatorname{cov}(\hat{\rho}_r, \hat{\rho}_s) \simeq \frac{1}{T}\sum_{j=-\infty}^{\infty} \{\rho_j\rho_{j+r-s} + \rho_{j-r}\rho_{j+s}\},$$

respectively. There are more references about the bias, the variance, and the covariance of the sample ACRF in Section 2.3.

The ACRF plays an important role in determining the order of an MA model. If $\{y_t\}$ is an MA(q) process, then the ACRF has a cutoff at lag q, i.e., $\rho_k = 0$ for $k = q+1, q+2, \ldots$, and the standard deviation estimate of the sample ACRF is

$$\hat{\sigma}(\hat{\rho}_k) \simeq \left\{\frac{1}{T}\left(1 + 2\sum_{j=1}^{q} \hat{\rho}_j^2\right)\right\}^{1/2}, \quad k = q+1, q+2, \ldots.$$

Because the bias and the standard deviation of the sample ACRF are $O(T^{-1})$ and $O(T^{-1/2})$, respectively, an approximate $100(1-\alpha)$ percent confidence interval of ρ_i is

$$\left(-z_{\alpha/2}\hat{\sigma}(\hat{\rho}_i),\ z_{\alpha/2}\hat{\sigma}(\hat{\rho}_i)\right), \quad i = q+1, q+2, \ldots,$$

where $z_{\alpha/2}$ is the z-value such that the area $\alpha/2$ lies to its right in the standard normal probability density. Because the ACRF of either an AR or a mixed ARMA process gradually tails off to 0, so does the sample ACRF. Thus, if q^* is the smallest natural number satisfying

$$\hat{\rho}_i \in \left(-z_{\alpha/2}\hat{\sigma}(\hat{\rho}_i),\ z_{\alpha/2}\hat{\sigma}(\hat{\rho}_i)\right), \quad i = q^*+1, q^*+2, \ldots,$$

then we regard the process as from an MA(q^*) model.

The coefficient $\phi_{k,k}$ is called the partial autocorrelation (also sometimes the reflection coefficient) at lag k, and $\{\phi_{k,k} \mid k = 1, 2, \ldots\}$ is called the partial autocorrelation function (PACF). It is another tool of Box and Jenkins' identification method. The term "partial autocorrelation" means that $\phi_{k,k}$ is the conditional correlation coefficient of y_t and y_{t+k} given $y_{t+1}, \ldots, y_{t+k-1}$. More precisely, we consider the Hilbert space of a stationary process $\{y_t\}$ with expectation as the inner product. Let $L_{s,t}$ be the subspace spanned by $\{y_{s+1}, \ldots, y_{t-1}\}$. If $\tilde{y}_s$ and $\tilde{y}_t$ are the projections of y_s and y_t onto $L_{s,t}$, respectively, then $\phi_{k,k}$ is the correlation of $y_s - \tilde{y}_s$ and $y_t - \tilde{y}_t$. Its proof can be found in standard textbooks like Hannan (1970a, pp. 21-22), T. W. Anderson (1971, p. 222), and Brockwell and Davis (1987, p. 164).

To utilize the PACF for model identification, it is necessary to study the asymptotic distribution of the YW estimate of the PACF.

Theorem 2.1. *The Asymptotic Distribution of the PACF*
Let $\{y_t\}$ be the AR(p) process. Then $\sqrt{T}\hat{\phi}_{p+1,p+1}$, $\sqrt{T}\hat{\phi}_{p+2,p+2}$, $\ldots$, are asymptotically independent random variables with means 0 and variances 1. □

There are more references about Theorem 2.1 and other characteristics of the PACF in Section 2.3.

If the underlying process is from an AR(p) model, then $\phi_{p+1,p+1} = \phi_{p+2,p+2} = \cdots = 0$. Because the bias of $\{\hat{\phi}_{k,k}\}$ is $O(T^{-1})$, Theorem 2.1 implies an approximate $100(1-\alpha)$ percent confidence interval of $\phi_{k,k}$ is

$$\left(-z_{\alpha/2}\frac{1}{\sqrt{T}},\ z_{\alpha/2}\frac{1}{\sqrt{T}}\right), \quad k = p+1, p+2, \ldots.$$

On the other hand, the PACF of either an MA or a mixed ARMA process tends to 0 gradually, hence so does the YW estimate of the PACF. Thus, if p^* is the smallest integer satisfying

$$\hat{\phi}_{k,k} \in \left(-z_{\alpha/2}\frac{1}{\sqrt{T}},\ z_{\alpha/2}\frac{1}{\sqrt{T}}\right), \quad k = p^*+1, p^*+2, \ldots,$$

then the realization can be regarded as from an AR(p^*) model.

Box-Jenkins' method is not very useful for identifying an ARMA model when neither p nor q is 0. Its ACRF consists of a mixture of damped exponentials and/or damped sine waves after the first $q-p$ lags, and the PACF consists of a mixture of damped exponentials and/or damped sine waves after the first $p-q$ lags. Thus, even if the graphs of $\{\rho_k\}$ and $\{\phi_{k,k}\}$ are error free, the simple inspection of their graphs would not, in general, yield unique values of p and q. The difficulty is compounded when their estimates are substituted for the true functions.

There is an earlier work by Rudra (1952), who used the ACRF and the PACF to choose the orders of pure AR and pure MA processes. He applied Bartlett's formula for the asymptotic variance of the sample ACRF. To test the cutoff property of the PACF, he recommended the use of the likelihood ratio criterion

$$\frac{\sqrt{T}\hat{\phi}_{k,k}}{\sqrt{1-\hat{\phi}_{k,k}^2}},$$

which is asymptotically normally distributed with mean 0 and variance 1 for $k = p+1, p+2, \ldots$. Kendall, Stuart, and Ord (1983, p. 637) have warned that we should be careful not to place too great a reliance on Box-Jenkins' method, for successive autocorrelations tend to be highly correlated. Moreover, if T is small, the biases of the sample ACRF and the YW estimate of the PACF should be counted.

2.2 The Inverse Autocorrelation Method

2.2.1 Inverse Autocorrelation Function

The inverse autocorrelation function (IACF) is the ACRF associated with the reciprocal of the spectral density of a time series. Consider a stationary time series with spectral density $S(\lambda)$. We define the inverse spectral density by

$$Si(\lambda) = \frac{1}{4\pi^2 S(\lambda)}, \quad -\pi \leq \lambda \leq \pi.$$

If $S(\lambda) \neq 0$ for $-\pi \leq \lambda \leq \pi$, then $Si(\lambda)$ has the Fourier expansion

$$Si(\lambda) = \frac{1}{2\pi} \sum_{j=-\infty}^{\infty} \sigma i(j) \exp(-i\lambda j),$$

where

$$\sum_{-\infty}^{\infty} | \sigma i(j) | < \infty.$$

(See, e.g., Davis [1966, p.105].) Applying the inverse Fourier transform to $Si(\lambda)$ yields

$$\sigma i(k) = \int_{-\pi}^{\pi} Si(\lambda) \exp(i\lambda k) d\lambda, \quad k = 0, \pm 1, \pm 2, \ldots,$$

which is called the inverse autocovariance at lag k. The inverse autocorrelation at lag k is defined by

$$\rho i(k) = \frac{\sigma i(k)}{\sigma i(0)}, \quad k = 0, \pm 1, \pm 2, \ldots.$$

Then, $\{\sigma i(k)\}$ and $\{\rho i(k)\}$ are called the inverse autocovariance function and the inverse autocorrelation function (IACF), respectively.

The notion of the IACF has been employed in time series analysis for a long time; for instance, the IACF has been used in a linear interpolation problem. (See, e.g., Grenander and Rosenblatt [1957, pp. 82-85] and Battaglia [1983].) Let $\{x_t \mid t = \ldots, -1, 0, 1, \ldots\}$ be a stationary process. Assume that the entire time series has been observed except at time t. The expectation of the mean squared error linear interpolator

$$E\left(x_t - \sum_{j \neq 0} \beta_j x_{t-j}\right)^2$$

has the minimum if $\beta_j = \rho i(j)$ for $j = 0, \pm 1, \pm 2, \ldots$. It is also known that the IACF has an orthogonal relation with the ACRF, i.e.,

$$\sum_{j=-\infty}^{\infty} \rho_j \rho i(k+j) \neq 0 \quad \text{if } k \neq 0.$$

It indicates that the infinite-dimensional inverse autocorrelation matrix is the inverse of the infinite-dimensional autocorrelation matrix. This property has been used to find an approximate inverse of the covariance matrix by Whittle (1952b), Shaman (1975, 1976), and Bhansali (1983b, 1990).

Let $\{y_t\}$ be from the ARMA(p, q) model (1.1). Then its spectral density and inverse spectral density are

$$S(\lambda) = \frac{\sigma^2 \mid \theta(\exp(i\lambda)) \mid^2}{2\pi \mid \phi(\exp(i\lambda)) \mid^2},$$
$$Si(\lambda) = \frac{1}{2\pi\sigma^2} \frac{\mid \phi(\exp(i\lambda)) \mid^2}{\mid \theta(\exp(i\lambda)) \mid^2}.$$

Because the invertibility implies $S(\lambda) \neq 0$ on $[-\pi, \pi]$, the IACF is

$$\rho i(k) = \int_{-\pi}^{\pi} \frac{\mid \phi(\exp(i\lambda)) \mid^2}{\mid \theta(\exp(i\lambda)) \mid^2} \exp(i\lambda k) d\lambda \Big/ \int_{-\pi}^{\pi} \frac{\mid \phi(\exp(i\lambda)) \mid^2}{\mid \theta(\exp(i\lambda)) \mid^2} d\lambda,$$

which is the autocorrelation at lag k of the ARMA(q, p) model

$$\theta(B)y_t = \phi(B)v_t.$$

This manifests the duality theorem that the IACF of the ARMA(p, q) process equals the ACRF of the dual ARMA(q, p) process. Therefore, all the properties of the ACRF of the dual ARMA process can be used to interpret the IACF. For example, if $q \neq 0$, then the IACF satisfies the inverse extended Yule-Walker (IEYW) equations

$$\rho i(k) = \theta_1 \rho i(k-1) + \cdots + \theta_q \rho i(k-q), \quad k = p+1, p+2, \ldots,$$

and then $\rho i(k)$ attenuates as $k \to \infty$. If $q = 0$, then the IACF becomes

$$\rho i(k) = \begin{cases} \left(-\phi_k + \sum_{j=1}^{p-k} \phi_j \phi_{j+k}\right) / \left(1 + \sum_{j=1}^{p} \phi_j^2\right), & k = 1, \ldots, p \\ 0, & k > p. \end{cases}$$

Thus, we can see the similarity between the PACF and the IACF. McLeod (1984) has discussed applications of the duality theorem to the ARMA model and to a multiplicative seasonal ARMA model.

2.2.2 Estimates of the Spectral Density

A natural estimate of the spectral density is the periodogram defined by

$$I(\lambda) = \frac{1}{2\pi} \sum_{j=1-T}^{T-1} \hat{\sigma}(j) \exp(-i\lambda j),$$

which equals

$$I(\lambda) = \frac{1}{2\pi T} \mid \sum_{t=1}^{T} (y_t - \bar{y}) \exp(-i\lambda t) \mid^2 .$$

Schuster (1898, 1899, 1906a, 1906b) has introduced the periodogram to search for unknown periodicities of empirical data such as Wolfer's sunspot data. It is well-known that the periodogram is inconsistent and that $I(\lambda_1)$ and $I(\lambda_2)$ are asymptotically independent for $\lambda_1 \neq \lambda_2$. Because the inconsistency and the asymptotic independence result in erratic and unreliable behavior of the periodogram, it is necessary to modify the periodogram to obtain good estimates of the spectral density. There are two kinds of spectral density estimates based on the periodogram. One is of the form

$$\hat{S}(\lambda) = \frac{1}{2\pi} \sum_{j=-K_T}^{K_T} w(j)\hat{\sigma}(j) \exp(-i\lambda j),$$

where $w(j)$ and K_T are called the lag window and the truncation point, respectively. (See, e.g., Blackman and Tukey [1958].) It is called a weighted covariance estimate. The other is of the form

$$\hat{S}(\lambda) = \int_{-\pi}^{\pi} W(\theta) I(\lambda - \theta) d\theta,$$

where $W(\theta)$ is called the spectral window. It is called a smoothed periodogram estimate and is based on a comment by Daniell (see Bartlett [1946, pp. 88-90]) that the covariance of $I(\lambda_1)$ and $I(\lambda_2)$ is so small that the variance of $I(\lambda)$ taken over a broad band of frequencies can be made as small as that of the sample ACVF.

If

$$w(j) = \int_{-\pi}^{\pi} W(\theta) \exp(-i\theta j) d\theta, \quad | j | \leq K_T - 1,$$

then the convolution theorem yields that a weighted covariance estimate is equivalent to a smoothed periodogram estimate. These estimates are called the traditional spectral density estimates. (See, e.g., Kay and Marple [1981].) Because the behavior of $\hat{S}(\lambda)$ depends on the window, we should choose a nice window to obtain a good traditional spectral density estimate. Also, it should be noted that the lag window $w(j)$ will have to depend on λ. Some comparisons of different windows can be found in Neave (1972), Harris (1978), and Brillinger (1981, pp. 52-60).

Using a digital computer, we can calculate $\hat{S}(\lambda)$ only for a finite number of λ's. To calculate the traditional spectral density estimates, we use a fast Fourier transform (FFT) algorithm, which was rediscovered by Cooley and Tukey (1965) and Gentleman and Sande (1966). For more details about FFT, consult the references in Section 2.3.

The window estimates have some shortcomings like leakages, side-lobes, limited bandwidths, and even negative-valued estimates. Moreover, the use of windows is a violation of a basic rule called the *principle of data reduction* by Ables (1974), i.e., the result of any transformation imposed on the experimental data shall incorporate and be consistent with all relevant data and be maximally noncommittal with regard to unavailable data. The window estimates violate this principle in two ways: the unavailable data are assumed and the available data are distorted by windows. Some analysts have tried to find more appropriate methods of estimating spectral densities. One of them is a parametric method using the AR model, which was proposed by Parzen (1968, 1969) and Akaike (1969b). Burg (1967, 1968) has given a concrete rationale to the AR model method that the spectral density maximizing the (differential) entropy rate subject to the first $s+1$ autocovariance constraints is the spectral density of the AR(s) process satisfying the constraints. It is called Burg's theorem.

To explain Burg's theorem in detail, we first consider a particle problem. Let V be any subspace of R^T and let $\boldsymbol{x}$ be a point of the subspace. Denote the volume of a convex set including $\boldsymbol{x}$ by $\Delta v = \Delta v(\boldsymbol{x})$, and let $n(\Delta v)$ be the number of particles (such as molecules) in the convex set. Then, the probability density function $f(\boldsymbol{x})$ of the particles is defined as

$$f(\boldsymbol{x}) = \lim_{\Delta v \to 0} \frac{n(\Delta v)}{\Delta v}.$$

In statistical mechanics a system is said to be in equilibrium if the probability density function is the closest to the uniform probability density function over the space V among the ones satisfying some physical constraints. An example of the constraint is the average energy of the particle is a constant c. If we denote the energy of the particle at $\boldsymbol{x}$ by $e(\boldsymbol{x})$, then the constraint is formulated as

$$\int e(\boldsymbol{x}) f(\boldsymbol{x}) d\boldsymbol{v} = c.$$

In order to choose the probability density function which is the closest to the uniform probability density function subject to some constraints, there should be a reasonable measure of distance of a probability density function from the uniform probability density function. In statistical mechanics the entropy is used as such a measure. More specifically, the entropy of a joint probability density function $f(\boldsymbol{x})$ of a random vector $\boldsymbol{x}$ is defined by

$$H(\boldsymbol{x}) = H(f) = -\int f(\boldsymbol{x}) \ln f(\boldsymbol{x}) d\boldsymbol{x},$$

which is also regarded as a measure of uncertainty or of ignorance. The particle problem is to choose the probability density function $f(\boldsymbol{x})$ maximizing $H(\boldsymbol{x})$ subject to the energy constraint For more details about the notion of entropy readers may refer to Khinchin (1956) and Ellis (1985).

Jaynes (1957-1982) has proposed the *principle of maximum entropy* as follows. When we choose a probability density function subject to some constraints, we should select the one that maximizes the entropy among the ones satisfying the constraints. The reason is the maximum entropy probability density function is considered as the one that can be realized in the greatest number of ways (Jaynes [1968, p. 231]). Under certain constraints, the maximum entropy probability density functions belong to the exponential family. (See, e.g., Kagan, Linnik, and Rao [1973, pp. 408-410].)

In time series analysis the principle of maximum entropy has been primarily used to estimate spectral densities. Let $\{y_t\}$ and $S(\lambda)$ be a stationary stochastic process and its spectral density, respectively. The entropy rate h is defined by

$$h = \lim_{T \to \infty} \frac{1}{T} H(y_1, \ldots, y_T).$$

Szegö's theorem in Grenander and Szegö (1955, pp. 63-66) yields

$$h = \frac{1}{2} \ln(2\pi e) + \frac{1}{4\pi} \int_{-\pi}^{\pi} \ln\{2\pi S(\lambda)\} d\lambda.$$

Burg has claimed that the spectral density maximizing the entropy rate h subject to the autocovariance constraints $\sigma(0) = \alpha_0, \ldots, \sigma(s) = \alpha_s$ is

$$S_M(\lambda) = \frac{\sigma_s^2}{2\pi} \frac{1}{| 1 - \phi_{s,1} \exp(-i\lambda) - \cdots - \phi_{s,s} \exp(-i\lambda s) |^2},$$

where $\phi_{s,1}, \ldots, \phi_{s,s}$ and σ_s^2 are the YW solutions of the constraints, i.e.,

$$\sum_{l=1}^{s} \phi_{s,l} \alpha_{|j-l|} = \alpha_j, \quad j = 1, \ldots, s,$$

$$\sigma_s^2 = - \sum_{l=0}^{s} \phi_{s,l} \alpha_l, \quad \phi_{s,0} = -1.$$

It is clear that $S_M(\lambda)$ is the spectral density of the AR(s) model. Various proofs of Burg's theorem have been proposed as referred to in Section 2.3.

To estimate a spectral density by the AR method, we first decide the order s of the AR model, and then obtain the estimates $\hat{\phi}_{s,1}, \ldots, \hat{\phi}_{s,s}$ $\hat{\sigma}_s^2$. The spectral estimate is

$$\hat{S}_M(\lambda) = \frac{\hat{\sigma}_s^2}{2\pi} \frac{1}{| 1 - \hat{\phi}_{s,1} \exp(-i\lambda) - \cdots - \hat{\phi}_{s,s} \exp(-i\lambda s) |^2}.$$

We call it the AR spectral density estimate or the maximum entropy spectral density estimate. For detailed properties of the maximum entropy spectral density estimate, consult the references in Section 2.3.

Some comparisons between the window method and the maximum entropy method have been presented. Lacoss (1971) and Kaveh and Cooper (1976) have shown that the maximum entropy estimate has a much higher resolution than any of the window estimates, particularly when the data record is of short duration. Beamish and Priestley (1981) have done an intensive simulation study and made the following conclusions.

- The maximum entropy method tends to gain when the underlying model can be easily approximated by AR models, and the window method gains when the opposite holds.
- A strong advantage of the maximum entropy method is that there are some objective tools for choosing the optimal AR order such as the FPE, the AIC, and the CAT, which will be discussed in Chapter 3. On the other hand, we should subjectively choose the window and the truncation point to estimate the spectral density by the window method. However, it should be mentioned that the AR orders selected by the objective tools may be unsatisfactory estimates.
- Even though both the maximum entropy and the window estimates may distort the true spectral density, the effects of smoothing and

side-lobe formation can often be ascertained from the window estimates, whereas the distortion by the maximum entropy method is incapable of similar determination.

Other spectral density estimation methods have been presented such as the method using ARMA models (see, e.g., Cadzow [1982] and Zhang and Taketa [1987]), Pisarenko's harmonic decomposition method (1972, 1973) and the maximum likelihood method (see, e.g., Capon [1969] and Lacoss [1971]). There have been a lot of advances in spectral analysis over the last two decades. Interested readers may consult the references in Section 2.3.

2.2.3 Estimates of the IACF

We may estimate the inverse autocovariance function as

$$\widehat{\sigma i}(k) = \int_{-\pi}^{\pi} \frac{1}{4\pi^2 \hat{S}(\lambda)} \exp(i\lambda k) d\lambda.$$

If $S(\lambda)$ is calculated at $N_T + 1$ points equally spaced in $[0, \pi]$, then

$$\widehat{\sigma i}(k) = \frac{1}{4\pi N_T} \sum_{j=1}^{2N_T} \frac{1}{\hat{S}(\frac{\pi j}{N_T})} \exp\left(ik\frac{\pi j}{N_T}\right), \quad k = 0, \ldots, N_T.$$

The corresponding IACF estimate is

$$\widehat{\rho i}(k) = \frac{\widehat{\sigma i}(k)}{\widehat{\sigma i}(0)}.$$

If the spectral density is estimated by a window method with window $\{w(j)\}$ and truncation point K_T, then a smoothed function $k(\cdot)$ can be defined by

$$k\left(\frac{j}{K_T}\right) = \begin{cases} w(j), & j = 0, \cdots, K_T \\ 0, & \text{otherwise.} \end{cases}$$

If the sequence $\{K_T\}$ is chosen so that $K_T \to \infty$ and $K_T/T \to 0$ as $T \to \infty$, then the IACF estimate $\{\widehat{\rho i}_w(k)\}$ has the following asymptotic properties (Bhansali [1980]). If some regularity conditions on the time series, on the function $k(\cdot)$, and on the sequence $\{K_T\}$ are satisfied, then $\widehat{\rho i}_w(k)$ is a consistent estimator of $\rho i(k)$. Also, under some regularity conditions,

$$\sqrt{\frac{2TN_T}{K_T}}\left(\widehat{\rho i}_w(1) - \rho i(1)\right), \ldots, \sqrt{\frac{2TN_T}{K_T}}\left(\widehat{\rho i}_w(m) - \rho i(m)\right)$$

are asymptotically normally distributed with means 0 and covariances

$$\lim_{T\to\infty} \frac{2TN_T}{K_T} \operatorname{cov}\left(\widehat{\rho i}_w(k), \widehat{\rho i}_w(j)\right)$$
$$\frac{4\pi}{\sigma i^2(0)} \int_{-\pi}^{\pi} \{\cos(\lambda k) - \rho i(k)\}\{\cos(\lambda j) - \rho i(j)\} Si^2(\lambda) d\lambda \int_{\infty}^{\infty} k^2(x) dx.$$

Let $\{\widehat{\rho i}_s(k)\}$ be the maximum entropy spectral density using an AR(s) model. If the sequence of AR orders is chosen so that $s \to \infty$ and $s^3/T \to 0$ as $T \to \infty$, then $\{\widehat{\rho i}_s(k)\}$ has the following asymptotic properties (Bhansali [1980]). Under some regularity conditions, $\widehat{\rho i}_s(k)$ is a consistent estimate of $\rho i(k)$. Also, if some regularity conditions are satisfied, then, as $T \to \infty$,

$$\sqrt{T}\left(\widehat{\rho i}_s(1) - \rho i(1)\right), \ldots, \sqrt{T}\left(\widehat{\rho i}_s(m) - \rho i(m)\right)$$

are asymptotically normally distributed with means 0 and covariances

$$\lim_{T\to\infty} T \operatorname{cov}\left(\widehat{\rho i}_s(k), \widehat{\rho i}_s(j)\right)$$
$$= \frac{4\pi}{\sigma i^2(0)} \int_{-\pi}^{\pi} \{\cos(\lambda k) - \rho i(k)\}\{\cos(\lambda j) - \rho i(j)\} Si^2(\lambda) d\lambda.$$

Bhansali (1983c) has examined the window and the maximum entropy estimates of the IACF through a simulation study and has concluded the following. The variance of $\widehat{\rho i}_s(k)$ tends to be smaller than that of $\widehat{\rho i}_w(k)$. The bias of $\widehat{\rho i}_s(k)$ is smaller than that of $\widehat{\rho i}_w(k)$ if the underlying time series can be easily approximated by an AR model. Otherwise, $\widehat{\rho i}_w(k)$ has a smaller bias. Thus, if the purpose of estimating the IACF is to determine the orders of an ARMA process, then the choice between the window and the maximum entropy methods may depend upon whether or not the time series can be approximated easily by an AR model. If this information is not available, the choice may be based on other considerations such as personal taste and the availability of the necessary computer programs.

Hipel, McLeod, and Lennox (1977), McLeod, Hipel, and Lennox (1977), and Chatfield (1979) have preferred the maximum entropy method to the window one, because the former can be done in a computationally easier way without frequency domain concepts. The invertibility assumption implies an ARMA(p, q) process can be represented by an AR(∞) model

$$y_t = \sum_{j=1}^{\infty} \pi_j y_{t-j} + v_t.$$

It can be well-approximated by an AR(s) model

$$y_t = \sum_{j=1}^{s} \pi_j y_{t-j} + v_t,$$

where s is sufficiently large. Because the IACF of the AR(s) process is the same as the ACRF of its dual MA(s) process, the IACF of the AR(s) process is

$$\rho i(k) = \begin{cases} \left(-\pi_k + \sum_{j=1}^{s-k} \pi_j \pi_{j+k}\right) / \left(1 + \sum_{j=1}^{s} \pi_j^2\right), & k = 1, \ldots, s, \\ 0, & k = s+1, s+2, \ldots. \end{cases}$$

In order to estimate the IACF using the approximate AR(s) model, we first estimate $\{\pi_j\}$ by solving the sample YW equations

$$\hat{\sigma}(k) = \sum_{j=1}^{s} \hat{\pi}_j \hat{\sigma}(k-j), \quad k = 1, \ldots, s,$$

and then substitute the AR estimate $\hat{\pi}_j$ for π_j.

Hosking (1980a) has derived the following asymptotic distribution of the sample IACF of an ARMA(p, q) process using the dual ARMA(q, p) model and Theorem 1.2. If $\{y_t\}$ is from the ARMA(p, q) model (1.1), then

$$\sqrt{T}\left(\widehat{\rho i}(1) - \rho i(1)\right), \ldots, \sqrt{T}\left(\widehat{\rho i}(m) - \rho i(m)\right)$$

are asymptotically normally distributed with means 0 and covariances

$$\begin{aligned} &\lim_{T \to \infty} \operatorname{cov}\{\sqrt{T}\left(\widehat{\rho i}(r) - \rho i(r)\right), \sqrt{T}\left(\widehat{\rho i}(s) - \rho i(s)\right)\} \\ &= \sum_{j=-\infty}^{\infty} \{\rho i(j+r)\rho i(j+s) + \rho i(j-r)\rho i(j+s) + 2\rho i(r)\rho i(s)\rho i^2(j) \\ &\qquad -2\rho i(r)\rho i(j)\rho i(j+s) - 2\rho i(s)\rho i(j)\rho i(j+r)\}. \end{aligned}$$

This result may be extended to any stationary stochastic process satisfying Anderson and Walker's assumptions (1964), for such a process can be approximated by an ARMA model with sufficiently high orders. For more IACF estimation methods, consult the references in Section 2.3.

2.2.4 Identification Using the IACF

Cleveland (1972) has proposed to use the IACF as well as the ACRF for ARMA model identification. If the underlying process is from an AR(p) model, its IACF has a cutoff at lag p, i.e., $\rho i(k) = 0$ for $k = p+1, p+2, \ldots$. Also, the asymptotic bias and the asymptotic standard deviation of the estimate of the IACF are $O(T^{-1})$ and $O(T^{-1/2})$, respectively. On the other hand, the IACF of a pure MA or a mixed ARMA process attenuates gradually as the lag increases, and so does its estimate. Thus, if p^* is the smallest integer such that $\widehat{\rho i}(k)$ is in the confidence interval of $\rho i(k)$ for $k = p^* + 1, p^* + 2, \ldots$, then the underlying process is regarded as an

AR(p^*) process. Consequently, the IACF plays the same role as the PACF in Box-Jenkins' identification procedure. Cleveland's method is not useful for identifying mixed ARMA models by the same reason as Box-Jenkins'. It should be noted that the covariance of $\widehat{\rho i}(k_1)$ and $\widehat{\rho i}(k_2)$ may be non-negligible compared with the variance of $\widehat{\rho i}(k)$. Thus, the selection rule based on the simultaneous confidence region should be applied.

Some statisticians prefer the IACF to the PACF in ARMA model identification. One of the reasons is that no new body of knowledge regarding the relationship between a time series and its IACF needs to be developed due to the duality between the IACF and the ACRF of ARMA models. Moreover, if $\phi_k = 0$ in an AR process, then $\rho i(k)$ should not be far from 0. Therefore, it is possible to detect the lags at which the AR coefficients are zeros by examining the IACF estimate. The latter property has been illustrated by Cleveland (1972) and McLeod, Hipel, and Lennox (1977). Based on the above characteristics and some simulation results, Chatfield (1979) and Bhansali (1983c) have asserted that the IACF is generally superior to the PACF in identifying an AR model.

However, more calculations are necessary to estimate the IACF than the PACF. The confidence interval of the IACF depends heavily upon the estimation method. Although it is known that the asymptotic bias of $\widehat{\rho i}(k)$ is $O(T^{-1})$, its refined form has not been derived yet. The asymptotic variance of the IACF estimate is more complicated than that of the PACF estimate, which is T^{-1}. Therefore, there is more statistical error when estimating a confidence interval of the IACF than of the PACF. Abraham and Ledolter (1984) have reported simulation results showing that the IACF is less powerful than the PACF and explained the reason as follows. In the case of an AR(p) process we can easily show that

$$\mid \phi_{p,p} \mid \geq \mid \rho i(p) \mid = \frac{\mid -\phi_p \mid}{\sum_{j=0}^{p} \phi_j^2}.$$

Because $\phi_{k,k} = \rho i(k) = 0$ for $k = p+1, p+2, \ldots$, the information from the IACF estimate could suggest the AR($p-1$) model more frequently than the PACF estimate. The situation will become much worse when the variances are taken into account because

$$\operatorname{var}\left(\widehat{\rho i}(k)\right) \approx \frac{1}{T}\left\{1 + 2\sum_{j=1}^{s} \rho i^2(j)\right\} > \frac{1}{T} \approx \operatorname{var}(\hat{\phi}_{k,k}),$$

for $k = p+1, p+2, \ldots$. This may be explained via Kanto's theorem (1984). Consider the Hilbert space L spanned by finite random terms $y_t, \ldots, y_{t+j}$ with expectation as the inner product, and let L_j be the subspace spanned by $y_{t+1}, \ldots, y_{t+j-1}$. If $\check{y}_t$ and $\check{y}_{t+j}$ are the projections of y_t and y_{t+j} onto L_j, respectively, then $\rho i(j)$ is the correlation of $y_t - \check{y}_t$ and $y_{t+j} - \check{y}_{t+j}$. Because $y_t - \check{y}_t$ and $y_{t+p} - \check{y}_{t+p}$ are the projections to $L - L_p$, $\rho i(p)$ is closer to 0 than $\phi_{p,p}$ in the case of AR processes.

McClave (1978b) and Bhansali (1983a) have utilized the IACF with some penalty functions, which will be discussed in Chapter 3, to identify MA processes. By analogy with the IACF, Hipel, McLeod, and Lennox (1977) have defined the inverse partial autocorrelation function (IPACF) of an ARMA(p, q) process by the PACF of its dual ARMA(q, p) process. Abraham and Ledolter (1984) have preferred the IPACF to the ACRF in detecting the orders of pure MA processes. Bhansali (1983b) has discussed the IPACF in detail.

2.3 Additional References

- (Section 2.1) For details about the bias, the variance, and the covariance of the sample ACRF, refer to Marriott and Pope (1954) and Kendall (1954).
- (Section 2.1) Some characteristics of the PACF have been discussed by Ramsey (1974) and Hamilton and Watts (1978). Theorem 2.1 was proven by Dixon (1944) and Quenouille (1949a, 1949b). Another simple proof can be found in Choi (1990b). The multivariate version of Theorem 2.1 was presented by Bartlett and Rajalakshman (1953), Morf, Vieira, and Kailath (1978), and Sakai (1981). Yajima (1985) investigated the asymptotic properties of the sample ACVF and of the PACF estimate of a multiplicative ARIMA model.
- (Section 2.2) For FFT algorithms, readers may refer to Runge (1903), Brigham and Morrow (1967), Cooley, Lewis, and Welch (1967, 1970a, 1970b), Bergland (1969), Brigham (1974), Silverman (1977), Brillinger (1981, pp. 64-66), Robinson (1982), Elliot and Rao (1982), and the references therein.
- (Section 2.2) Many have tried to prove Burg's theorem. Smylie, Clarke, and Ulrych (1973, pp. 402-419) have established it by variational methods. Edward and Fitelson (1973) have proved it using the Lagrange multiplier method. Burg (1975), Ulrych and Bishop (1975), Haykin and Kesler (1979, pp. 16-21), and Robinson (1982) have followed Smylie et al.'s method. Ulrych and Ooe (1979) and McDonough (1979) have used Edward et al.'s proof. Van den Bos (1971) has tried to maximize the entropy h subject to the autocovariance constraints by differential calculus. But further argument is required to complete the proof. Also, refer to Feder and Weinstein (1984). In 1977, Akaike proved it using Kolmogorov's identity (see Priestley [1981, pp. 604-606]). Grandell, Hamrud, and Toll (1980) have used the same idea to prove it. Choi and Cover (1984, 1987) have presented an information theoretic proof, which needs neither the stationarity assumption nor the Gaussian assumption. Choi (1986b) has proved it using

Hadamard's inequality and Choi (1991d) has proved it using the LU decomposition method. The last two proofs extend the maximum entropy spectrum problem to a probability density function problem subject to the autocovariance constraints.

- (Section 2.2) Properties of the maximum entropy spectral density have been studied by many time series analysts including Kromer (1969), Berk (1974), and Ensor and Newton (1988). Recently, Parzen (1983a) has reviewed the AR spectral density in detail. A multivariate maximum entropy spectral density has been given by Choi (1991e). Franke (1985b) has shown that an ARMA(p, q) process has the maximum entropy subject to the $p+1$ autocovariances and the q impulse responses. Also, refer to Ihara (1984).

- (Section 2.2) For recent developments in spectral analysis, readers may refer to the books edited by Childers (1978), Haykin (1979), Haykin and Cadzow (1982), Brillinger and Krishnaiah (1983), and Kesler (1986). Even though it is a little out of date, the survey paper by Kay and Marple (1981) is good enough to know what is going on with spectral analysis. Also, refer to Gutowski, Robinson, and Treitel (1978), Nitzberg (1979), Thomson (1981), Fuhrmann and Liu (1986), Kay and Shaw (1988), and the references therein.

- (Section 2.2) There are other IACF estimation methods presented by Battaglia (1983, 1986, 1988), Kanto (1987), and Subba Rao and Gabr (1989).

3

Penalty Function Methods

Since the early 1970s, some estimation-type identification procedures have been proposed. They are to choose the orders k and i minimizing

$$P(k,i) = \ln \check{\sigma}_{k,i}^2 + (k+i)\frac{C(T)}{T},$$

where $\check{\sigma}_{k,i}^2$ is an estimate of the white noise variance obtained by fitting the ARMA(k,i) model to the observations. Because $\check{\sigma}_{k,i}^2$ decreases as the orders increase, it cannot be a good criterion to choose the orders minimizing it. If the orders increase, the bias of the estimated model will decrease while the variance increases. Therefore, we should compromise between them. For this purpose we add the penalty term, $(k+i)C(T)/T$, into the model selection criterion $P(k,i)$. The *penalty function identification methods* are regarded as objective.

It is natural to assume that the selected orders increase as T does. In practice, no upper bound of the orders k and i might be prescribed because the minimum of $P(k,i)$ is expected to be found for quite small values of k and i. However, some upper bounds will be necessary to prove theorems. It means an uncritical minimization of $P(k,i)$ over too large a range of values of its arguments could lead to trouble (see, e.g., Hannan and Deistler [1988, pp. 162-164]). Consider the AR model identification. When the AR(k) model is fitted to a T-realization $\{y_1,\ldots,y_T\}$ by the YW procedure, the estimate of the white noise variance is

$$\hat{\sigma}_k^2 = \inf_{\phi_1,\ldots,\phi_k} \frac{1}{2\pi}\int_{-\pi}^{\pi} I(\lambda) \mid \sum_{l=0}^{k} \phi_l \exp(il\lambda) \mid^2 d\lambda, \quad \phi_0 = -1,$$

where

$$I(\lambda) = \frac{1}{T} \mid \sum_{t=1}^{T} y_t \exp(it\lambda) \mid^2 .$$

As discussed in Section 2.2, Szegö's theorem implies

$$\inf_k \hat{\sigma}_k^2 = \exp\left\{\frac{1}{2\pi}\int_{-\pi}^{\pi} \ln I(\lambda) d\lambda\right\}.$$

An (1982) has shown that

$$\lim_{T\to\infty} \inf_k \hat{\sigma}_k^2 = \sigma^2 \exp(-\gamma), \text{ a.s.},$$

where γ is Euler's constant. Therefore, if the order k is greatly overstated, then the estimate $\hat{\sigma}_k^2$ will be biased downward. Thus, we would rather impose some upper bounds for k and i. Let K and I be upper bounds of k and i, respectively. If it is necessary to symbolize that they increase as T does, then we denote them as K_T and I_T. Even though there is no specific guideline on how to choose the upper bounds, Hannan's setting,

$$K_T \le (\ln T)^a, \; I_T \le (\ln T)^a \quad (0 < a < \infty),$$

is considered as reasonable. There are more references about this choice in Section 3.9.

3.1 The Final Prediction Error Method

Consider a T-realization $\{y_1, \ldots, y_T\}$ of the AR(p) process satisfying

$$y_t = \phi_1 y_{t-1} + \cdots + \phi_p y_{t-p} + v_t.$$

Let $\{x_t\}$ be another AR(p) process that is independent of $\{y_t\}$ but having the same statistical structure as $\{y_t\}$, i.e.,

$$x_t = \phi_1 x_{t-1} + \cdots + \phi_p x_{t-p} + u_t,$$

where the white noise process $\{u_t\}$ has the same distribution as $\{v_t\}$, but the two white noise processes are independent of each other. Then, a natural predictor of x_{T+1} is

$$\hat{x}_{T+1} = \hat{\phi}_{p,1} x_T + \cdots + \hat{\phi}_{p,p} x_{T-p+1},$$

where $\hat{\phi}_{p,1}, \ldots, \hat{\phi}_{p,p}$ are the YW estimates obtained by fitting the AR(p) model to the observations $\{y_1, \ldots, y_T\}$. Using Theorem 1.4 and the independence of $\{x_t\}$ and $\{\hat{\phi}_{p,1}, \ldots, \hat{\phi}_{p,p}\}$, Akaike (1969a) has shown that, for large T, the mean square prediction error is

$$E(\hat{x}_{T+1} - x_{T+1})^2 \simeq \sigma^2 \left(1 + \frac{p}{T}\right).$$

The consistency of the sample ACVF and Theorem 1.4 yield

$$E(\hat{\sigma}_p^2) \simeq \left(1 - \frac{p}{T}\right) \sigma^2$$

for large T. Thus, $(1 - p/T)^{-1}\hat{\sigma}_p^2$ is a less biased estimate of σ^2 than $\hat{\sigma}_p^2$. The corresponding estimate of the asymptotic mean square error of $\hat{x}_{T+1}$ is

$$(1 + p/T)(1 - p/T)^{-1}\hat{\sigma}_p^2,$$

which is asymptotically equivalent to $(1+2p/T)\hat{\sigma}_p^2$. Akaike has defined the final prediction error (FPE)

$$\mathrm{FPE}(p) = \left(1 + 2\frac{p}{T}\right)\hat{\sigma}_p^2,$$

which is an asymptotically unbiased and consistent estimator of the mean square error of the one-step ahead predictor $\hat{x}_{T+1}$. As Akaike (1970a, p. 209) has mentioned, the FPE should be replaced by

$$\mathrm{FPE}(p) = \hat{\sigma}_p^2\left(1 + 2\frac{p+1}{T}\right),$$

if the mean of the process is unknown. Because it is reasonable to use the predictor that minimizes the mean square prediction error, it is meaningful to estimate the order by minimizing the FPE(k). The order estimate will be called the minimum FPE estimate (MFPEE).

The FPE(k) is composed of two parts. The first part, $\hat{\sigma}_k^2$, corresponds to the variance of the best linear predictor for a given k, and the other part, $(2k/T)\hat{\sigma}_k^2$, is due to statistical deviations of $\{\hat{\phi}_{k,1},\ldots,\hat{\phi}_{k,k}\}$ from the true AR parameters. In general, the former decreases and the latter increases for fixed T, as k increases. Thus, the minimization of the FPE(k) is a compromise to obtain an optimal order. If an unnecessarily large value of k is adopted, then the FPE(k) tends to be large. If k is less than the true order of the process, then the bias of $\hat{\sigma}_k^2$ is large. Thus, the method of choosing k that minimizes the FPE(k) balances the risks between the bias of a low order and the variance of a high order.

Historically speaking, Mallows' C_p (1973), a method similar to the FPE, was proposed to select suitable independent variables in regression analysis. In time series analysis Davisson (1965) has presented the asymptotic mean square error of predicting one-step ahead. Its refined form by Yamamoto (1976) and Fuller and Hasza (1981) is

$$E(\hat{x}_{T+1} - x_{T+1})^2 = \sigma^2\left(1 + \frac{k}{T}\right) + O\left(\frac{1}{T^{3/2}}\right).$$

Davisson then has proposed to choose the value of k minimizing an estimate of $\sigma^2(1+k/T)$ as an optimal order of the process. Davisson's method has practical difficulty because of the choice of σ^2 estimate. However, Davisson, based on simulation results, made the following comment:

> If a pessimistic result is desired rather than the apparently optimistic approximation of $\sigma^2(1+p/T)$, then $\sigma^2(1+2p/T)$ seems to be a good choice. In fact, more exhaustive empirical results have verified this for $T > 10$ and $p/T < 0.5$.

Davisson (1966) applied the procedure to analyzing signal plus noise data. Akaike (1974a) mentioned that he had proposed the FPE procedure without knowledge of Davisson's paper. Nowadays the procedure is called

Akaike's FPE criterion because it has been popular since his publications. The FPE procedure has been applied to more general statistical modeling than AR order selection. For a more complete perspective on the FPE method, consult the references in Section 3.9.

In practice it is impossible to calculate the FPE for all orders. Usually an integer K is preassigned as an upper bound of possible orders. Clearly, the integer K should be large enough to contain the true order. Several subjective benchmarks about the upper bound K have been proposed. For example, Akaike (1970a) has recommended that $K = T/10$ should be a reasonable choice for ordinary observation size T. Ulrych and Bishop (1975) have recommended that K should be less than $T/2$. As discussed before, Hannan and Deistler (1988) have imposed the upper bound $(\ln T)^a$ $(1 < a < \infty)$.

The FPE procedure has a tendency to overestimate the AR order shown by many time series analysts such as Akaike (1970a), Kashyap (1980), and Hannan (1982).

Theorem 3.1 If $\{y_t\}$ is from an AR(p) model, then

$$\lim_{T\to\infty} P\{\text{FPE}(k) > \text{FPE}(p)\} = 1, \quad k < p,$$

$$\lim_{T\to\infty} P\{\text{FPE}(k) > \text{FPE}(p)\} = P\{\chi^2_{k-p} > 2(k-p)\}, k \geq p,$$

where χ^2_r is the chi-squared random variable with r degrees of freedom. □

The inconsistency of the MFPEE has been a matter of controversy. Based on his simulation results, Bhansali (1973) asserted, "When the true order of an AR process is not known, the FPE procedure tends to overfit the model." Akaike (1974a) pointed out that Bhansali's disappointing result was due to his incorrect definition of the related statistic: Bhansali used $\hat{\sigma}^2_k$ instead of $(1+k/T)\hat{\sigma}^2_k$ as the estimate of σ^2. Jones (1975) supported Akaike's opinion by resimulating Bhansali's experiments. Akaike has commented that the inconsistency does not necessarily mean a serious drawback of the FPE procedure because the probability of adopting $\hat{p}$ larger than the true order p is not intolerable. The consistency problem will be discussed more in Section 3.7.

Davisson (1965) and Bhansali (1974) commented on the possibility of using the multi-step ahead predictor in ARMA model identification. Piccolo and Tunnicliff-Wilson (1984) proposed to use the covariance matrix of multi-step ahead predictors for ARMA model identification. However, no identification procedure based on multi-step ahead predictions has become popular so far. In general, an optimal multi-step ahead predictor is not necessarily determined by an optimal one-step ahead predictor. Findley (1984) has suggested that we should use a more direct approach rather than a fitted model to produce the forecasts if the goal is to do m-step ahead forecasting ($m > 1$). In other words, to forecast y_{T+m} ($m > 1$) using the observations $y_1, \ldots, y_T$, it is better to minimize

$$\sum_{t=p}^{T-m} \left\{ y_{t+m} - \sum_{j=1}^{p} \phi_j y_{t-j+1} \right\}^2$$

than to estimate the AR parameters from the observations and then to predict y_{T+m} using the estimated AR model recursively. There are some references about the asymptotic mean square error of a multi-step ahead predictor in Section 3.9.

3.2 Akaike's Information Criterion

3.2.1 Kullback-Leibler Information Number

Kullback and Leibler (1951) have proposed an information theoretic quantity to measure how much different a probability density function $g(x)$ is from another probability density function $f(x)$ in a sense of statistical distinguishability:

$$I(f; g) = \int f(x) \ln \left(\frac{f(x)}{g(x)} \right) dx.$$

If the integral exists, the quantity $I(f; g)$ is called the Kullback-Leibler information number of f relative to g. Many statistical theorems have shown that it is a natural measure to discriminate between two probability density functions.

In statistical hypothesis testing, the logarithm of the power of the likelihood ratio test can be represented by the Kullback-Leibler information number. Let $\{x_j\}$ be a sequence of independent and identically distributed random variables with probability density function $p(x)$. Consider the hypothesis testing problem

$$H_0 : p = f \text{ versus } H_1 : p = g.$$

If

$$L_T = \prod_{t=1}^{T} \frac{g(x_t)}{f(x_t)},$$

$$\alpha_T = P(L_T \geq \lambda_T \mid H_0),$$

$$\beta_T = P(L_T < \lambda_T \mid H_1),$$

then the sequence of rejection regions $\{L_T \geq \lambda_T\}$ forms that of likelihood ratio tests with Type I error α_T and Type II error β_T. If $\alpha_T \to \alpha$ $(0 < \alpha < 1)$ as $T \to \infty$, then

$$\lim_{T\to\infty} \frac{1}{T} \ln \beta_T = -I(f;g).$$

This property has been known as Stein's lemma and has been shown by Chernoff (1952), Rao (1973, pp. 451-453), and Csiszár and Körner (1981, p. 28). Because it implies that the larger $I(f;g)$ the more information distinguishing $f(x)$ from $g(x)$, $I(f;g)$ is sometimes called the mean information for discrimination between the probability density functions f and g. This theorem also shows that $I(f;g)$ is non-negative.

Sanov (1957) has connected the Kullback-Leibler information number with the large deviation problem of random variables. Let x be a random variable with probability density function $g(x)$ and cumulative distribution function $G(x)$. Also, let $G_T(x)$ be its empirical distribution function after T independent trials. If $f(x)$ and $F(x)$ are another probability density function and its cumulative distribution function such that $I(f;g)$ exists, then

$$\lim_{\epsilon\to 0} \lim_{T\to\infty} \frac{1}{T} \ln P\left(\sup_x \mid G_T(x) - F(x) \mid < \epsilon\right) = -I(f;g).$$

Because $G_T(x)$ converges to $G(x)$ uniformly on the whole line as $T \to \infty$, Sanov's theorem says that the probability the supremum distance between $G(x)$ and $F(x)$ is less than ϵ is asymptotically $\exp\{-TI(f;g)\}$. Thus, if $F(x)$ and $G(x)$ are close in statistical senses, $I(f;g)$ should be small. There are more references about Sanov's theorem in Section 3.9.

Van Campenhout and Cover (1981) have linked it with conditional probability problems. Let $\{x_j\}$ be a sequence of independent and identically distributed random variables with probability density function $g(x)$ satisfying some regularity conditions. Then, the conditional probability density function of x_1 given the sample mean

$$\frac{1}{T}\sum_{j=1}^{T} x_j = \alpha$$

tends to the probability density function $f_\alpha(x)$ minimizing $I(f;g)$ among the probability density functions satisfying

$$\int x f(x) dx = \alpha.$$

There are more references about the conditional limit problem in Section 3.9.

From a decision-theoretic point of view, the Kullback-Leibler information number can be interpreted as a risk. If the loss function is

$$\ln\left\{\frac{f(x)}{g(x)}\right\},$$

then the risk becomes $I(f;g)$. (See, e.g., DeGroot [1970, p. 236].) Using the Taylor expansion it can be shown (see, e.g., Kendall and Stuart [1979, p. 446]) that $2I(f;g)$ is approximate to

$$\int \frac{\{f(x)-g(x)\}^2}{f(x)}dx,$$

which is similar to Pearson's X^2 statistic. Also, refer to Larimore (1983).

All of the above theorems show that the larger $I(f;g)$ is, the more different g is from f. Thus, the Kullback-Leibler information number is a good measure of statistical distances. Because an estimated probability density function $\check{f}(x)$ should be close to the true probability density function $f(x)$, it is reasonable to choose $\check{f}(x)$ minimizing the Kullback-Leibler information number $I(f;\check{f})$.

3.2.2 Akaike's Information Criterion

Akaike (1972a) has presented an identification method using the Kullback-Leibler information number as well as the consistency and the asymptotic normality of ML estimates. Let $x_1,\ldots,x_T$ be independent and identically distributed random variables with probability density function $f(x \mid \boldsymbol{\theta}^0)$, where $\boldsymbol{\theta}^0(\in R^K)$ is a fixed but unknown parameter vector. The estimation problem is to find $\boldsymbol{\theta}(\in R^K)$ so that $f(x \mid \boldsymbol{\theta})$ is close in some statistical senses to $f(x \mid \boldsymbol{\theta}^0)$. As discussed in the previous subsection, one of the reasonable estimation methods is to minimize the Kullback-Leibler information number,

$$I(\boldsymbol{\theta}^0;\boldsymbol{\theta}) = \int f(x \mid \boldsymbol{\theta}^0)\ln f(x \mid \boldsymbol{\theta}^0)dx - \int f(x \mid \boldsymbol{\theta}^0)\ln f(x \mid \boldsymbol{\theta})dx.$$

Let $R^k = \{(a_1,\cdots,a_k,0,\ldots,0)^t\}$ be a subspace of R^K for $k=1,\ldots,K$, and let $\tilde{\boldsymbol{\theta}}_k$ be the ML estimate of $\boldsymbol{\theta}^0$ restricted to R^k. Let J be the $K \times K$ Fisher information matrix. Then we can build a Hilbert space $\{\boldsymbol{\theta} \mid \boldsymbol{\theta} \in R^K\}$ with inner product operator and norm operator

$$(\boldsymbol{\alpha},\boldsymbol{\beta})_J = \boldsymbol{\alpha}^t J\boldsymbol{\beta}, \quad \parallel \boldsymbol{\alpha} \parallel_J = \sqrt{(\boldsymbol{\alpha},\boldsymbol{\alpha})_J}.$$

It is known (see, e.g., Kullback [1959]) that the Taylor expansion of $f(x \mid \tilde{\boldsymbol{\theta}}_k)$ at $\boldsymbol{\theta}=\boldsymbol{\theta}^0$ yields

$$\begin{aligned} I(\boldsymbol{\theta}^0;\tilde{\boldsymbol{\theta}}_k) &\simeq \frac{1}{2}\parallel \tilde{\boldsymbol{\theta}}_k - \boldsymbol{\theta}^0 \parallel_J^2 \\ &= \frac{1}{2}\parallel \tilde{\boldsymbol{\theta}}_k - \boldsymbol{\theta}_k^0 \parallel_J^2 + \frac{1}{2}\parallel \boldsymbol{\theta}_k^0 - \boldsymbol{\theta}^0 \parallel_J^2, \end{aligned}$$

where $\boldsymbol{\theta}_k^0$ is the projection of $\boldsymbol{\theta}^0$ onto R^k. The asymptotic normality of the

ML estimates implies

$$E\{I(\boldsymbol{\theta}^0; \tilde{\boldsymbol{\theta}}_k)\} \simeq \frac{1}{2} \parallel \boldsymbol{\theta}_k^0 - \boldsymbol{\theta}^0 \parallel_J^2 + \frac{k}{2T}.$$

Let $l(k, K)$ be the log-likelihood ratio statistic for testing hypotheses

$$H_0 \ : \ \boldsymbol{\theta}^0 \in R^k \ \text{ versus } \ H_1 : \boldsymbol{\theta}^0 \in R^K - R^k.$$

Cochran's theorem and the asymptotic normality of the ML estimates imply the asymptotic distribution of $(-2)l(k, K)$ is a noncentral χ^2 distribution with $K - k$ degrees of freedom and noncentrality parameter $T \parallel \boldsymbol{\theta}_k^0 - \boldsymbol{\theta}^0 \parallel_J^2$. Thus,

$$\frac{1}{2T}\{(-2)l(k, K) + 2k - K\}$$

is an asymptotically unbiased estimate of $E\{I(\boldsymbol{\theta}^0; \tilde{\boldsymbol{\theta}}_k)\}$, and it equals

$$\frac{1}{2T}\left\{2\sum_{t=1}^{T} \ln f(x_t \mid \tilde{\boldsymbol{\theta}}_K) - K\right\} + \frac{1}{2T}\left\{-2\sum_{t=1}^{T} \ln f(x_t \mid \tilde{\boldsymbol{\theta}}_k) + 2k\right\}.$$

Because the first term does not depend on k, the order minimizing the estimate of $E\{I(\boldsymbol{\theta}^0; \tilde{\boldsymbol{\theta}}_k)\}$ equals the one minimizing

$$\text{AIC}^*(k) = -2\sum_{t=1}^{T} \ln f(x_t \mid \tilde{\boldsymbol{\theta}}_k) + 2k.$$

This is called Akaike's information criterion (AIC), and the order minimizing the AIC is the minimum AIC estimate (MAICE). The MAICE has been used to select optimal models in various fields of statistics including time series analysis. For details, consult the references in Section 3.9.

The AIC procedure may be one of the most distinguished novelties in statistical modeling. We cite **this week's citation classic** of *Current Contents: Engineering, Technology and Applied Sciences* (December 21, 1981, p. 22).[1]

> Akaike's paper (1974a), "A new look at the statistical model identification," describes how the problem of statistical model selection can systematically be handled by using an information criterion (AIC) introduced by the author in 1971. The basic idea underlying the introduction of the criterion is explained and its practical utility is demonstrated by numerical examples. The *Science Citation Index* and the *Social Science Citation Index* indicate that this paper has been cited over 180 times since 1974.

---

[1]The present author thanks to Eric Thurschwell at the Institute for Scientific Information for sending a copy of this part of the *Current Contents*.

The derivation of the AIC* has been limited to the sequence of independent and identically distributed random variables. Akaike (1973c) has commented that the same line of discussion can be extended to cover the case of finite order Markov processes by Billingsley's approach (1961, pp. 14-16). Following the suggestion, Ogata (1980) has derived the AIC procedure for AR models. Recently, Hurvich and Tsai (1989) presented a refined estimate of the expected Kullback-Leibler information number for AR processes and proposed to use a modified AIC

$$\mathrm{AIC}(k) = \ln \tilde{\sigma}_k^2 + \frac{T+k}{T-k-2}.$$

The AIC for ARMA processes is defined as

$$\mathrm{AIC}(k,i) = \ln \tilde{\sigma}_{k,i}^2 + \frac{2}{T}(k+i),$$

where $\tilde{\sigma}_{k,i}^2$ is the ML estimate of the white noise variance as defined before. The AIC is a special case of the AIC* divided by T. Findley (1985) has derived the asymptotic bias of the maximum log-likelihood as an estimate of the expected log-likelihood of an ARMA model and has concluded that the AIC procedure applied to Gaussian ARMA processes is justifiable.

Ozaki (1977) has applied the AIC procedure to the data sets of Box and Jenkins (1976) and has concluded that the minimum AIC method overcomes many difficulties of Box-Jenkins' identification procedure. For mixed ARMA(p,q) processes with large p and q, Kitagawa (1977) proposed a quick heuristic way of searching optimal orders based on AIC values and residual variances.

3.3 Generalizations

In this section, only AR models will be considered. We denote the true order of an AR process by p and let K be the largest value among the possible orders.

The MFPEE is asymptotically equivalent to the MAICE because

$$\ln \mathrm{FPE}(k) = \mathrm{AIC}(k) + o\left(\frac{1}{T}\right).$$

Thus, only the asymptotic properties of the MAICE will be mentioned in this section.

Shibata (1976) has presented the asymptotic distribution of the MAICE.

Theorem 3.2. Let $\{Z_j\}$ be a sequence of independent and identically distributed χ_1^2 random variables. Let

$$a(0) = b(0) = 1.$$

For $j = 1, 2, \ldots$, let

$$a(j) = P\left(\sum_{m=1}^{n}(Z_m - 2) > 0,\ n = 1, \ldots, j\right),$$

$$b(j) = P\left(\sum_{m=1}^{n}(Z_m - 2) \leq 0,\ n = 1, \ldots, j\right).$$

Then, the asymptotic distribution of the MAICE $\hat{p}$ is

$$\lim_{T\to\infty} P(\hat{p} = k) = \begin{cases} a(k-p)b(K-k), & p \leq k \leq K \\ 0, & \text{otherwise.} \end{cases} \quad \square$$

It should be kept in mind that the asymptotic distribution of the MAICE depends not on the AR coefficients but on K as shown in Theorem 3.2. Shibata (1976) has presented a table of numerical values of the asymptotic distribution when $K = 10$. It shows that the asymptotic probability of selecting the true order is over 70 percent and that of choosing $p+1$ is over 11 percent. Moreover, the probability of selecting too high an order dies out fairly quickly. For more details, refer to Shibata (1976, Table 2) and Duong (1984, Tables IV and VI). Hannan (1980b) has generalized Shibata's theorem to the case of ARMA(k, i) models for $k = p$ and $i \geq q$ or for $k \geq p$ and $i = q$. In this case the results of Woodroofe (1982) imply that the probability of overfitting is never larger than 0.288. We will discuss it in Section 3.7. There are more references about the asymptotic distribution of the MAICE in Section 3.9.

The inconsistency raised some objections to the FPE and the AIC procedures. (See, e.g., Kendall, Stuart, and Ord [1983, p. 637].) In contrast, Findley (1984) said, "An important consequence of the usually approximate nature of model is that, for model selection, consistency can be an undesirable property." Also, Hannan and Quinn (1979) have remarked that the concept of consistency loses some of its significance in a situation where we consider the selected AR$(\hat{p})$ model as an optimal approximation to an AR(∞) process.

To circumvent the inconsistency problem, the FPE and the AIC have been modified by

$$\text{FPE}_\alpha(k) = \left(1 + \alpha\frac{k}{T}\right)\tilde{\sigma}_k^2,$$

$$\text{AIC}_\alpha(k) = \ln\tilde{\sigma}_k^2 + \alpha\frac{k}{T}.$$

The estimates minimizing the FPE_α and AIC_α are called the $\text{MFPE}_\alpha\text{E}$ and $\text{MAIC}_\alpha\text{E}$, respectively.

Bhansali and Downham (1977) have presented the asymptotic distribution of the MAIC$_\alpha$E, which is a generalized result of Theorem 3.2.

Theorem 3.3. Let $\{Z_j\}$ be a sequence of independent and identically distributed χ_1^2 random variables. Let

$$a_\alpha(0) = b_\alpha(0) = 1.$$

For $j = 1, 2, \ldots$, let

$$a_\alpha(j) = P\left(\sum_{m=1}^{n}(Z_m - \alpha) > 0,\ n = 1, \ldots, j\right),$$

$$b_\alpha(j) = P\left(\sum_{m=1}^{n}(Z_m - \alpha) \leq 0,\ n = 1, \ldots, j\right).$$

Then, the asymptotic distribution of the MAIC$_\alpha$E $\hat{p}_\alpha$ is

$$\lim_{T\to\infty} P(\hat{p}_\alpha = k) = \begin{cases} a_\alpha(k-p)b_\alpha(K-k), & p \leq k \leq K \\ 0, & \text{otherwise.} \end{cases} \quad \square$$

Theorem 3.3 shows that, for any fixed α (> 0), the MAIC$_\alpha$E is still inconsistent. However, we can penalize overparameterization more rigidly by increasing α beyond 2.

Theorem 3.4. For $\alpha > 1$, let

$$h(\alpha) = \alpha^{\frac{1}{2}} \exp\left(-\frac{\alpha - 1}{2}\right).$$

If $\alpha > 1$, then, under the condition of Theorem 3.3,

$$\lim_{T\to\infty} P(\hat{p}_\alpha = p) > 1 - h(\alpha),$$

$$\lim_{T\to\infty} P(\hat{p}_\alpha = k) < h(\alpha)^{k-p}, \quad p \leq k \leq K. \quad \square$$

Theorem 3.4 means that if α is sufficiently large, then the probability of selecting the true order can be made arbitrarily close to 1. For fixed α (> 1), the probability of selecting an order larger than the true order decreases exponentially as k increases. Thus, the choice of the upper bound of K is not so crucial as far as $p \leq K$. Based on simulation results, Bhansali and Downham (1977) have concluded as follows.

- If $T \geq 300$, the empirical distributions are close to the asymptotic distribution in Theorem 3.4.

- There is a considerable advantage in choosing $\alpha = 4$ rather than $\alpha = 2$.

- If T is small, and if the dependence of y_t and y_{t-p} is strong, then the frequency of fitting the true order is increased by choosing α greater than 2.

- If T is small, and if the dependence of y_t and y_{t-p} is weak, then the frequency of fitting the true order may not necessarily increase by choosing α greater than 2.

Because the true order and the strength of the dependence between y_t and y_{t-p} are unknown, it is impossible to obtain a simple rule to select α. Bhansali (1979, p. 206) has suggested that several values of $\alpha \in (2, \ln T)$ should be tried and has commented that it may be necessary to select $\alpha \in (1, 2)$ when T is less than 100.

Shibata (1980, 1981a, 1981b) and Taniguchi (1980) have shown that the MAIC$_\alpha$E (or equivalently, the MFPE$_\alpha$E) is asymptotically efficient when $\alpha = 2$. More precisely, when loss functions appropriate for prediction and spectral density estimation are used, the MAIC$_\alpha$E with $\alpha \neq 2$ can lead to relatively large loss compared with the MAICE. Shibata (1983) has concluded that when the MAICE is larger than 10, one may trust its optimality in the sense of the loss functions as well as in the Kullback-Leibler information sense even while taking its overfitting behavior into consideration. Shibata (1984) has pointed out that if the MAICE is not large, then a minimax choice of α and the corresponding MAIC$_\alpha$E procedure seem to be more reasonable.

Again it should be noted that in many situations an ARMA model selected from a finite number of observations cannot describe the true model perfectly. Duong (1984) has proposed a method to choose several AR models as candidates for the selected model as follows. For any positive constant c, let $R_\alpha(c)$ be a subset of $\{0, 1, \ldots, K\}$ as

$$R_\alpha(c) = \{k \mid \mathrm{AIC}_\alpha(k) - c < \mathrm{AIC}_\alpha(\hat{p}_\alpha)\}.$$

It was shown that

$$\lim_{T\to\infty} P\{p \in R_\alpha(c)\} = \sum \prod_{j=1}^{K-p} \frac{1}{\mu_j!} \left\{ \frac{1 - P(\chi_j^2 > \alpha + c)}{j} \right\}^{\mu_j!},$$

where the summation extends over all $(K-p)$-tuples $(\mu_1, \ldots, \mu_{K-p})$ of non-negative integers satisfying $\mu_1 + 2\mu_2 + \cdots + (K-p)\mu_{K-p} = K - p$. For a given probability p^*, let

$$c^* = c^*(p^*, \alpha, T) = \min\{c \mid P[p \in R_\alpha(c)] > p^*\}.$$

Then $R_\alpha(c^*)$ is a collection of AR orders, to which the true order p belongs with probability p^*. It bears analogy to a confidence interval and is particularly useful when insufficiency of data kept us from making a crystally clear choice. The asymptotic value c^* can be derived when $T \to \infty$. He tabulated the limiting values for $\alpha = 1, 2, 3, 4$ and $p^* = 0.90,\ 0.95,\ 0.99$.

Bhansali (1988) applied the FPE_α to order determination of AR processes with infinite variance.

3.4 Parzen's Method

If all the roots of $\theta(z) = 0$ are outside the unit circle, there exists the unique infinite order polynomial

$$\pi(z) = 1 - \pi_1 z - \pi_2 z^2 - \cdots,$$

which converges to $\phi(z)\theta^{-1}(z)$ on $(-1, 1)$. Then, the ARMA(p, q) process $\{y_t\}$ can be represented by the AR(∞) model satisfying

$$\pi(B) y_t = v_t.$$

The function $\pi(z)$ is called the infinite autoregressive transfer function.

To select a model for a T-realization $\{y_1, \ldots, y_T\}$ of a stationary time series, we usually postulate that the realization is from an ARMA model. Then the orders and the parameters are to be estimated from the realization. However, Parzen's approach (1974) is somewhat different. The true model is assumed to have an AR(∞) representation, and the transfer function $\pi(z)$ is estimated instead of the orders and the parameters. Thus, the selected order $\hat{p}$ is interpreted not as an estimate of a finite AR order but as providing an optimal finite order AR approximation to the true AR(∞) process. The selected order $\hat{p}$ will be treated as a function of T and will be assumed to tend to ∞ as T does. This idea was exploited by Yule (1927) and Whittle (1952a).

To estimate $\pi(z)$, Parzen (1975) has used the penalty function $E\{J(\breve{\pi})\}$ of an estimate $\breve{\pi}(z)$, where

$$J(\breve{\pi}) = \frac{1}{2\pi} \int_{-\pi}^{\pi} \left\{ \frac{\breve{\pi}(e^{i\lambda}) - \pi(e^{i\lambda})}{\pi(e^{i\lambda})} \right\}^2 d\lambda.$$

If $\breve{\pi}$ minimizes the penalty function, then the corresponding spectral density estimate will be a good one. Because there are only a finite number of observations, it is natural to approximate the AR(∞) process by an AR model with finite order. Define the YW estimate of $\pi(z)$ by

$$\hat{\pi}_k(z) = 1 - \hat{\pi}_{k,1} z - \cdots - \hat{\pi}_{k,k} z^k,$$

where $\hat{\pi}_{k,1}, \ldots, \hat{\pi}_{k,k}$ are the YW estimates. Using the asymptotic properties presented by Kromer (1969), it can be shown

$$E\{J(\hat{\pi}_k)\} = 1 - \frac{\sigma^2}{\sigma_k^2} + \frac{k}{T},$$

where σ_k^2 is the same as the one defined in Section 1.3. To select an optimal estimate of $\pi(z)$ among $\{\hat{\pi}_k(z) \mid k = 1, 2, \ldots\}$, we choose k to minimize

$$\text{CAT}^*(k) = 1 - \frac{(T-k)\hat{\sigma}^2}{T\hat{\sigma}_k^2} + \frac{k}{T},$$

where

$$\hat{\sigma}^2 = (2\pi)\exp\left(\frac{1}{2\pi}\int_{-\pi}^{\pi} \ln \hat{S}(\lambda) d\lambda\right)$$

and $\hat{S}(\lambda)$ is a spectral density estimate. The CAT stands for the criterion of autoregressive transfer function.

The expectation $E\{J(\hat{\pi}_k)\}$ consists of two parts. The first part $(1 - \sigma^2/\sigma_k^2)$ represents the bias due to approximating the infinite order polynomial $\pi(z)$ to a finite order polynomial $\hat{\pi}_k(z)$. It has been known (Parzen [1969, Equation 5.10]) that the overall variance of the estimate $\hat{\pi}_k(z)$ is asymptotically equal to the second part k/T. Thus, the order minimizing the CAT* is a compromise between the bias and the variance.

Because

$$\frac{(T-k)\hat{\sigma}^2}{T\hat{\sigma}_k^2} = 1 + O\left(\frac{1}{T}\right),$$

it can be shown that

$$\text{CAT}^*(k) \simeq \text{AIC}(k) - \ln\hat{\sigma}^2 + O\left(\frac{1}{T}\right).$$

Thus, the CAT and the AIC procedures give asymptotically equivalent results. Bhansali (1986) has generalized it as

$$\text{CAT}_\alpha^*(k) = 1 - \frac{(T-k)\hat{\sigma}^2}{T\hat{\sigma}_k^2} + \alpha\frac{k}{T},$$

and has shown that the CAT_α^* and the AIC_α are asymptotically equivalent when $\alpha > 1$.

Unfortunately the statistic CAT* requires the estimate $\hat{\sigma}^2$, and then the spectral estimate $\hat{S}(\lambda)$. To avoid estimating them, Parzen (1977) has proposed to minimize J_k instead of $E\{J(\breve{\pi})\}$, where

$$J_k = \frac{1}{2\pi}\int_{-\pi}^{\pi}\left[\left\{\frac{1}{\breve{\sigma}_p^2}\breve{\pi}(e^{i\lambda}) - \frac{1}{\sigma^2}\pi(e^{i\lambda})\right\} / \left\{\frac{1}{\sigma^2}\pi(e^{i\lambda})\right\}\right]^2 d\lambda.$$

Using recursive relations in filtering theory (see, e.g., Kailath [1974]), Parzen has derived the asymptotic equation

$$J_k = \frac{1}{T}\sum_{j=1}^{k}\frac{1}{\sigma_j^2} + \frac{1}{\sigma^2} - \frac{1}{\sigma_k^2}.$$

Therefore it is reasonable to choose the order minimizing

$$\mathrm{CAT}(k) = \begin{cases} -\left(1+\frac{1}{T}\right), & k=0 \\ \frac{1}{T}\sum_{j=1}^{k}\hat{\sigma}_j^{-2} - \hat{\sigma}_k^{-2}, & k=1,2,\ldots. \end{cases}$$

The corresponding estimate is called the minimum CAT estimate (MCATE). There are some references having illustrative examples of the CAT procedure in Section 3.9.

Tong (1979) has proposed to use a modification of the CAT as

$$\mathrm{CAT}_*(k) = \frac{1}{T}\sum_{j=0}^{k}\frac{1}{\breve{\sigma}_j^2} - \frac{1}{\breve{\sigma}_k^2},$$

where

$$\breve{\sigma}_j^2 = \frac{T}{T-j}\hat{\sigma}_j^2, \quad j=0,1,\ldots,$$

which is an unbiased estimate σ_j^2. Using it, he has shown an asymptotic local equivalence of the MAICE and the MCATE. We know

$$\mathrm{CAT}_*(k) - \mathrm{CAT}_*(k-1) = \frac{1}{\breve{\sigma}_{k-1}^2}\left\{1-\left(1-\frac{1}{T}\right)\frac{1}{1+\triangle_k}\right\},$$

where

$$\triangle_k = \frac{\breve{\sigma}_k^2 - \breve{\sigma}_{k-1}^2}{\breve{\sigma}_{k-1}^2}.$$

Tong has used a slightly modified AIC

$$\mathrm{AIC}_*(k) = \ln\breve{\sigma}_k^2 + \frac{k}{T},$$

which satisfies

$$\begin{aligned} &\mathrm{AIC}_*(k) - \mathrm{AIC}_*(k-1) \\ &= \frac{1}{T} + \ln\frac{\breve{\sigma}_k^2}{\breve{\sigma}_{k-1}^2} \\ &= \frac{1}{T} + \ln(1+\triangle_k). \end{aligned}$$

Thus,

$$\begin{aligned}&\mathrm{CAT}_*(k) - \mathrm{CAT}_*(k-1)\\&\quad = \frac{1}{\breve{\sigma}^2_{k-1}} \{\mathrm{AIC}_*(k) - \mathrm{AIC}_*(k-1)\} + \left\{\frac{1}{T\breve{\sigma}^2_{k-1}}\triangle_k + O(\triangle_k^2)\right\}.\end{aligned}$$

Because the second term of the RHS is small compared to the first term, we can say that CAT(k) and AIC(k) rise and fall together as k changes. Therefore we can say that the local behavior of the CAT is asymptotically equivalent to that of the AIC, but we may not say that the MCATE and the MAICE are asymptotically equivalent because $\breve{\sigma}^2_{k-1}$ does depend on k. The present author has had some experiences that the CAT and the AIC procedures yield the same results, particularly when the underlying process is from a true AR model. As mentioned by Parzen (1983a, p. 239), it appears reassuring that quite different conceptual rudiments lead to similar conclusions in practice.

3.5 The Bayesian Information Criterion

The FPE, the AIC, and the CAT procedures have been widely used as objective criteria to determine orders in time series analysis, even though they result in inconsistent estimates of the true orders. As mentioned several times, it is hardly anticipated that a time series is from an exact ARMA model with fixed orders. Hence, we just try to choose the best approximate ARMA model for the given observations. Usually it is reasonable to assume that p and q increase as the data length increases because more data means more information. Thus, some say that we should not take the inconsistencies of the FPE, the AIC, and the CAT procedures into account too seriously in ARMA model identification. However, some disagree and propose new identification methods that result in consistent estimates of the true orders. Just after the possibility of using a Bayesian approach in model identification was mentioned by Akaike (1977b), some statisticians presented the Bayesian information criterion (BIC),

$$(-2)\ln(ML) + (\text{number of parameters}) \ln T.$$

We choose the order minimizing the BIC, which is called the minimum BIC estimate (MBICE). Especially for an ARMA process, the BIC is simplified as

$$\mathrm{BIC}(k,i) = \ln \tilde{\sigma}^2_{k,i} + (k+i)\frac{\ln T}{T},$$

where $\tilde{\sigma}^2_{k,i}$ is the ML estimate of the white noise variance.

3.5.1 Schwarz' Derivation

Schwarz (1978) derived the BIC for the Koopman-Darmois family.

Theorem 3.5. Let $x_1, \ldots, x_T$ be independent and identically distributed random variables with probability density function

$$f(x; \boldsymbol{\theta}) = \exp\{\boldsymbol{\theta}^t \mathbf{y}(x) - b(\boldsymbol{\theta})\}.$$

Here, $\boldsymbol{\theta}$ ranges over the parameter space Ω, which is a convex subset of the K-dimensional Euclidean space R^K. The other models have the parametric spaces $\omega_i = \Omega_i \bigcap \Omega$, where Ω_i is a k_i-dimensional linear submanifold of R^K. Let α_i be the prior probability that the ith model is the true one. Let μ_i be the conditional prior distribution of $\boldsymbol{\theta}$ on ω_i given the ith model, and assume that μ_i has a k_i-dimensional probability density function that is bounded and locally bounded away from 0 on ω_i. Then, the following holds.

1. The model with the largest posterior probability is the one maximizing
$$S(\bar{\mathbf{y}}, T, i) = \ln \int_{\omega_i} \alpha_i \exp[\{\boldsymbol{\theta}^t \bar{\mathbf{y}} - b(\boldsymbol{\theta})\}T] d\mu_i(\boldsymbol{\theta}),$$
where $\bar{\mathbf{y}} = \frac{1}{T}\sum_{t=1}^{T} \mathbf{y}(x_t)$.

2. For any i,
$$S(\bar{\mathbf{y}}, T, i) = T \sup_{\boldsymbol{\theta} \in \omega_i} \{\boldsymbol{\theta}^t \bar{\mathbf{y}} - b(\boldsymbol{\theta})\} - \frac{k_i}{2} \ln T + R,$$
where R is bounded in T. □

The first part of Theorem 3.5 can be proven using Bayes' formula, whereas the second part can be obtained through Schwarz' theorem (1971). Schwarz proposed using only the first two terms of $S(\bar{\mathbf{y}}, T, i)$ as the criterion. Because

$$T \sup_{\boldsymbol{\theta} \in \omega_i} \{(\boldsymbol{\theta}^t \bar{\mathbf{y}} - b(\boldsymbol{\theta})\}$$

is the maximum log-likelihood on ω_i, $(-2)S(\bar{\mathbf{y}}, T, i)$ is asymptotically equivalent to the BIC. Therefore, the order with the maximal posterior probability equals the MBICE asymptotically. It should be noted that the final result does not depend on the prior probabilities. G. C. Chow (1981) has questioned the use of only the first two terms of $S(\bar{\mathbf{y}}, T, i)$ as a model selection criterion and has commented that the Bayesian criterion will depend on the prior probability density if a refined approximation of $S(\bar{\mathbf{y}}, T, i)$ is used. Schwarz has remarked that if the assumptions in Theorem 3.5 are satisfied, the MAICE cannot be asymptotically optimal. Akaike (1978b) has made a counterargument by providing an asymptotic optimal property of the MAICE. Stone (1979) presented an interesting study related to these comments.

3.5.2 Kashyap's Derivation

Kashyap (1977) presented a Bayesian method for comparing different types of dynamic structures. The decision rule answers a wide variety of questions such as the advisability of a nonlinear transformation of data, limitations of a model yielding a perfect fit to the data, etc. Kashyap's other papers discussed it in association with AR processes (1978) and ARMA processes (1982) in detail.

Theorem 3.6. Let $\mathbf{y} = (y_1, \ldots, y_T)^t$ be an observation vector. Assume that there are r competing hypotheses $H_1, \ldots, H_r$, which are mutually exclusive so that any observation vector can come from only one of the r hypotheses. Let $p_i(\mathbf{y}_i; \boldsymbol{\alpha}_i)$ be the probability density function of H_i, where $\boldsymbol{\alpha}_i$ is a parameter vector belonging to $\omega_i(\subset R^{k_i})$ and has no zero-component. Denote the ML estimate of $\boldsymbol{\alpha}_i$ by $\tilde{\boldsymbol{\alpha}}_i$. If the prior probability density functions $f_i(\boldsymbol{\alpha}_i)$ and $p_i(\mathbf{y}_i; \boldsymbol{\alpha}_i)$ satisfy some regularity conditions, then the log of the posterior probability that H_i is the correct one for given $\mathbf{y}$ is

$$\ln P(H_i \mid \mathbf{y}) = \ln p_i(\mathbf{y}_i; \tilde{\boldsymbol{\alpha}}_i) - \frac{k_i}{2} \ln T + O_p(1). \quad \square$$

An optimal Bayesian decision rule of choosing the best model for a given time series is to assign $\mathbf{y}$ to H_k so that $P(H_k \mid \mathbf{y})$ is greater than or equal to $P(H_i \mid \mathbf{y})$ for $i = 1, \ldots, r$. Because the maximum log-likelihood is $O_p(T)$, $P(H_i \mid \mathbf{y})$ is asymptotically equivalent to

$$\ln p_i(\mathbf{y}_i; \tilde{\boldsymbol{\alpha}}_i) - \frac{k_i}{2} \ln T.$$

Thus, the criterion is asymptotically equivalent to BIC. Again it does not depend on the prior probability densities. However, the posterior probabilities may be influenced to a considerable degree by the prior probability densities when T is not so large. Two kinds of the priors were proposed for the ARMA case. Let $\boldsymbol{\alpha} = (\phi_1, \ldots, \phi_k, \theta_1, \ldots, \theta_i)^t$. If $\mathbf{y}$ is normally distributed, then a conjugate family consists of a multivariate normal probability density function for $\boldsymbol{\alpha}$ given σ^2 and a gamma probability density function for $(1/\sigma^2)$. (See, e.g., DeGroot [1970, pp. 177-180].) Kashyap (1977) used an improper density c/σ^2 for a positive constant c as a prior density of σ^2. Kashyap (1982) used also a uniform probability density function for $\boldsymbol{\alpha}$ and $(2\sigma^2)\exp(- \mid \ln \sigma^2 \mid)$ for σ^2.

3.5.3 Shortest Data Description

Rissanen (1978, 1979, 1980, 1982) proposed an identification procedure for ARMA processes using the *shortest data description principle.* Even though the idea is quite different from any Bayesian approach, the decision statistic is asymptotically equivalent to the BIC.

To introduce the shortest data description principle formally, we need a fundamental inequality of information theory.

Lemma 3.1. Let $\{p_j \mid j = 1, 2, \ldots\}$ and $\{q_j \mid j = 1, 2, \ldots\}$ be two sequences of non-negative terms satisfying

$$\sum_j p_j \geq \sum_j q_j,$$

where the summations are convergent. Then,

$$-\sum_j p_j \log_2 p_j \leq -\sum_j p_j \log_2 q_j,$$

where the equality holds *if and only if* $p_j = q_j$ for all j. □

This lemma is known as Gibb's theorem, which is a special case of Jensen's inequality. However, it can be easily shown from the fact that the Kullback-Leibler number is always non-negative. In order to interpret Gibb's theorem as a size inequality for the description length principle, we consider a random variable x taking on values $a_1, a_2, \ldots$ with probability mass function $P(x = a_j) = p_j$. Assume that codewords of length $n_1, n_2, \ldots$ are assigned to the symbols $a_1, a_2, \ldots$, respectively. The purpose is to construct a uniquely decipherable code that minimizes the average codeword length

$$n = \sum_j n_j p_j = -\sum_j p_j \log_2 2^{-n_j}.$$

The unique decipherability of the code implies that the codewords should satisfy the Kraft inequality

$$\sum_j 2^{-n_j} \leq 1.$$

(See, e.g., Rissanen [1983a].) Let $n_j = -\log_2 q_j$. Then

$$\sum_j q_j \leq 1 = \sum_j p_j.$$

Lemma 3.1 implies

$$-\sum_j p_j \log_2 p_j \leq -\sum_j p_j \log_2 q_j = n.$$

Therefore, the minimum of the average codeword length is

$$-\sum_j p_j \log_2 p_j,$$

which is the entropy of the random variable x *if and only if* $q_j = p_j$ for each j. (Also, refer to Ash [1965, p. 16].) In other words, the minimum cannot be achieved unless the model coincides with the true system.

Let $\{y_1, \ldots, y_T\}$ be a T-realization of the ARMA(p, q) model (1.1). Assume that $y_t = v_t = 0$ for $t \leq 0$. Define the parameter vector by

$$\boldsymbol{\xi} = (\xi_0, \xi_1, \ldots, \xi_{p+q})^t = (\sigma^2, \phi_1, \ldots, \phi_p, \theta_1, \ldots, \theta_q)^t.$$

As a criterion to determine the orders p and q, we adopt the shortest data description, i.e., the codeword length to describe $\mathbf{y} = (y_1, \ldots, y_T)^t$ should be minimized. Because $\mathbf{y}$ can be reconstructed from p, q, $\boldsymbol{\xi}$ and $\mathbf{v} = (v_1, \ldots, v_T)^t$, the codeword length of $\mathbf{y}$ can be written as

$$L(\mathbf{y}, p, q, \boldsymbol{\xi}) = L(\mathbf{v} \mid p, q, \boldsymbol{\xi}) + L(p, q, \boldsymbol{\xi}),$$

where $L(\mathbf{v} \mid p, q, \boldsymbol{\xi})$ is the conditional codeword length of $\mathbf{v}$ given p, q, and $\boldsymbol{\xi}$, and $L(p, q, \boldsymbol{\xi})$ is the codeword length of $(p, q, \boldsymbol{\xi})$. Assume that v_t is already written with precision $\pm\epsilon/2$ and that ξ_j, i.e., the $(j+1)$th element of $\boldsymbol{\xi}$, is written with precision $\pm\delta_j/2$ for $j = 0, \ldots, p+q$.

To obtain $L(\mathbf{v} \mid p, q, \boldsymbol{\xi})$, it is necessary to use Shannon's coding theorem (see, e.g., El Gamal and Cover [1980]), which is an application of the asymptotic equipartition property by Shannon (1948), McMillian (1953, pp. 207-209), and Breiman (1957).

Lemma 3.2. Let $\{x_1, \ldots, x_T\}$ be a sequence of independent and identically distributed random variables with range X and probability mass function $\{p_j \mid j \in X\}$. Let H be the entropy of each random variable, i.e.,

$$H = H(x_1) = -\sum_{j \in X} p_j \log_2 p_j.$$

Then, for any $\epsilon > 0$, there exist an integer T, an encoding rule f, and a decoding rule g such that

$$f : X^T \to \{2^0, 2^1, \ldots, 2^{T(H+\epsilon)}\},$$

$$g : \{2^0, 2^1, \ldots, 2^{T(H+\epsilon)}\} \to X^T,$$

$$P\{g(f(\boldsymbol{x})) \neq \boldsymbol{x}\} < \epsilon. \quad \square$$

Lemma 3.2 as well as Lemma 3.1 implies the average codeword length of $\boldsymbol{v}$ given p, q, and $\boldsymbol{\xi}$ is approximately TH bits. Assume that the white noise process $\{v_t\}$ has the normal probability density $\phi(v \mid 0, \sigma^2)$. Then,

$$P\left(\mid v_t - j\epsilon \mid < \frac{\epsilon}{2}\right) = \int_{(j-\frac{1}{2})\epsilon}^{(j+\frac{1}{2})\epsilon} \phi(z \mid 0, \sigma^2) dz \simeq \epsilon\phi(j\epsilon \mid 0, \sigma^2).$$

Thus, the law of large numbers yields an approximation

$$H \simeq -\frac{1}{T}\sum_{j=1}^{T}\log_2\{\epsilon\phi(v_t \mid 0, \sigma^2)\}.$$

Hence, the sequence $\{v_1, \ldots, v_T\}$ is described approximately by the following amount of bits:

$$L(\mathbf{v} \mid p, q, \boldsymbol{\xi}) = TH = \frac{1}{2}T\ \log_2\left(\frac{2\pi\sigma^2}{\epsilon^2}\right) + \frac{1}{2\ln 2}\sum_{t=1}^{T}\frac{v_t^2}{\sigma^2}.$$

To describe the orders p, q and the parameter vector $\boldsymbol{\xi}$, we need the following amount of bits:

$$L(p, q, \boldsymbol{\xi}) = \log_2(p+1)(q+1) + \sum_{j=0}^{p+q}\log_2\frac{\mid \xi_j \mid}{\delta_j}.$$

Combining them, we obtain a lemma about the codeword length to describe the realization $\{y_1, \ldots, y_T\}$ as follows.

Lemma 3.3. Assume that the ARMA(p, q) process is Gaussian and that v_t and ξ_j are written with precisions $\pm\epsilon/2$ and $\pm\delta_j/2$, respectively. Then, the codeword length to describe $\boldsymbol{y}$ is

$$\begin{aligned} L(\mathbf{y}, p, q, \boldsymbol{\xi}) &= \frac{1}{2}T\ \log_2\left(\frac{2\pi\sigma^2}{\epsilon^2}\right) + \frac{1}{2\ln 2}\sum_{t=1}^{T}\frac{v_t^2}{\sigma^2} \\ &\quad + \log_2(p+1)(q+1) + \sum_{j=0}^{p+q}\log_2\frac{\mid \xi_j \mid}{\delta_j}. \quad \square \end{aligned}$$

Our purpose is to find the vector $\boldsymbol{\xi}$ that minimizes the expected codeword length $E\{L(\mathbf{y}, p, q, \boldsymbol{\xi})\}$.

Theorem 3.7. Let $\tilde{\boldsymbol{\xi}} = (\tilde{\xi}_0, \ldots, \tilde{\xi}_{p+q})^t$ be the vector minimizing $L(\mathbf{y}, p, q, \boldsymbol{\xi})$ with fixed $\mathbf{y}$, p, and q. Let $\check{\xi}_j$ denote the number when $\tilde{\xi}_j$ is truncated to the level $\pm\delta_j/2$ for $j = 1, \ldots, p+q$. Assume that $\check{\xi}_1 - \tilde{\xi}_1, \ldots, \check{\xi}_{p+q} - \tilde{\xi}_{p+q}$ are independent and that $\check{\xi}_j - \tilde{\xi}_j$ has the uniform distribution on $[-\delta_j/2, \delta_j/2]$. Then

$$\begin{aligned} E\{L(\mathbf{y}, p, q, \boldsymbol{\xi})\} &= \frac{1}{2}\left\{T\ln\tilde{\xi}_0 + (p+q+1)\ln(T+2)\right\} \\ &\quad + \frac{1}{2}\left\{T + T\ln\frac{2\pi}{\epsilon^2} + 2\right\} + o(T). \quad \square \end{aligned}$$

Theorem 3.7 indicates that the optimal model based on the shortest data description principle is approximately the one minimizing the statistic

$$\mathrm{RC}(k,i) = \ln \tilde{\sigma}^2_{k,i} + (k+i+1)\frac{\ln(T+2)}{T},$$

where RC stands for Rissanen's criterion. Clearly the RC is asymptotically equivalent to the BIC, whenever the latter is applicable. The shortest data description principle, of which the described codeword length criterion is just one, has been applied to other modeling problems and, in fact, it has been taken as a fundamental principle for statistical inquiry. (See, for details, Rissanen [1983a, 1983b, 1984, 1987, 1989].)

3.5.4 Some Comments

The BIC was also derived for orthogonal polynomial regression by Akaike (1977a) and for the mean vector of a multivariate normal distribution by Akaike (1978c). Sawa (1978) derived it for distinguishing between alternative regression models and Leamer (1979) made a comment on Sawa's result. Poskitt and Tremayne (1983) studied it for a sufficiently general class of nonlinear time series models. Haggan and Oyetunji (1984) utilized it to select subset AR models. Pukkila (1989) applied it to identifying integrated AR models.

The consistency of the MBICE holds in many regression type models. Huang (1990b) has proved that the MBICE is strongly consistent for stationary, unstable, and explosive AR models. The consistency gives a reasonable interpretation that the MAICE is an upper bound of selected orders, and that the MBICE produces more consistent results in repeated application when the number of observations is large. However, Akaike (1976) has shown through maximum entropy spectral density estimation that the BIC procedure may underestimate the true orders. More precisely, the BIC procedure tends to result in an oversmoothed spectral density estimate compared with the MAICE when the true order of the underlying model is infinite. Akaike has commented that we should be careful not to ignore some of the possibly meaningful details of spectral density estimates unless the results obtained by various procedures are in good agreement. Stoica (1979) has presented a numerical comparison among the BIC procedure and other identification procedures and has concluded that the F-test and the BIC procedure give the best results. Neftci (1982) has applied the AIC and the BIC procedures to a group of major economic time series and has concluded that the MAICE turns out to be of substantially higher dimensions than the ones currently used in practice. In many articles Hannan has stressed that the inconsistency of the MAICE need not be a defect of the AIC method, for the observations are not generated by a true ARMA model and the estimated ARMA model may be fitted only as an approximation. For more details, consult the references in Section 3.9.

3.6 Hannan and Quinn's Criterion

Most of the works discussed in Sections 3.6 and 3.7 are done by E. J. Hannan and his colleagues. In these sections we employ their model defined as follows:

$$\phi(B)y_t = \theta(B)v_t, \tag{3.1}$$

where $\phi(B) = -\phi_0 - \phi_1 B - \cdots - \phi_p B^p$, $\theta(B) = -\theta_0 - \theta_1 B - \cdots - \theta_q B^q$, $\phi_0 = \theta_0 = -1$, $\phi_p \neq 0$, $\theta_q \neq 0$, and $\{v_t\}$ is a sequence of uncorrelated random variables with means 0 and variances σ^2 (> 0). Assume that all the roots of $\phi(z) = 0$ and $\theta(z)$ are outside the unit circle and they have no common root. It is also assumed that

$$E\{v_t \mid \mathcal{F}_{t-1}\} = 0, \ E\{v_t^2 \mid \mathcal{F}_{t-1}\} = \sigma^2, \ E\{v_t^4\} < \infty,$$

where $\mathcal{F}_t$ is the σ-algebra generated by $\{v_t, v_{t-1}, \ldots\}$ (or equivalently by $\{y_t, y_{t-1}, \ldots\}$). The first of these conditions is natural because the best linear prediction is suboptimal without it. It is equivalent to the requirement that the best linear predictor be the best among all predictors in the least squares sense. The second one could be considerably relaxed at the expense of an increase of complexity. For more details about these moment assumptions, refer to Hannan and Rissanen (1982, p. 82).

First, we consider only AR processes. Hannan and Quinn (1979) have provided an identification procedure minimizing

$$\ln \hat{\sigma}_k^2 + k\frac{C(T)}{T}$$

so that, as $T \to \infty$, the selected order is strongly consistent to the true order and $C(T)/T$ decreases as fast as possible.

Theorem 3.8. Let $\{y_t\}$ be generated by the AR(p) model satisfying the conditions discussed above. Let K be a predetermined upper bound satisfying $p \leq K$. If $\hat{p}$ is the order minimizing

$$\mathrm{HQC}(k) = \ln \hat{\sigma}_k^2 + 2ck\frac{\ln \ln T}{T}, \quad c > 1,$$

subject to $k \leq K$, then $\hat{p}$ is strongly consistent to p. □

The HQC stands for Hannan and Quinn's criterion. The order minimizing the HQC is called the minimum HQC estimate (MHQCE).

The outline of Hannan and Quinn's proof of Theorem 3.8 is as follows. The Levinson-Durbin algorithm yields

$$\mathrm{HQC}(k) = \ln \hat{\sigma}(0) + \sum_{j=1}^{k} \ln\left(1 - \hat{\phi}_{k,k}^2\right) + 2kc\frac{\ln \ln T}{T}.$$

Thus, HQC(k) increases at k by

$$\ln\left(1-\hat{\phi}_{k,k}^2\right)+2c\frac{\ln\ln T}{T}.$$

Because $\hat{\phi}_{p,p}$ converges to $\phi_p(\neq 0)$,

$$\ln\left(1-\hat{\phi}_{p,p}^2\right)+2c\frac{\ln\ln T}{T}<0, \text{ a.s.},$$

for sufficiently large T. Thus, HQC(k) cannot have the asymptotic minimum if $k<p$. Hannan and Quinn have shown that if $k>p$, then

$$\hat{\phi}_{k,k}=b_k(T)\left(\frac{2\ln\ln T}{T}\right)^{1/2},$$

where the sequence $\{b_k(T)\mid T=1,2,\ldots\}$ has its limit points in the interval $[-1,1]$. This property results in Theorem 3.8.

Hannan (1980b, p. 1072) has commented that Theorem 3.8 may hold for $c=1$ but this would require more delicate analysis. In practical problems c is usually taken to be 1. However, if HQC(k) is replaced by

$$\ln\hat{\sigma}_k^2+k\frac{C(T)}{T},$$

where $C(T)$ satisfies

$$\lim_{T\to\infty}C(T)=\infty,\quad \lim_{T\to\infty}\frac{C(T)}{\ln\ln T}<2,$$

then the strong consistency of the resulting estimate cannot hold. The reason is

$$\ln\left(1-\hat{\phi}_{k,k}^2\right)+\frac{C(T)}{T}$$

will be negative infinitely often with respect to T for $k>p$.

Hannan and Quinn (1979, p. 195) have made comments on the HQC procedure as follows. If T is large and an AR model is thought to be a good approximation to the true structure, then the use of the MHQCE would have something to recommend it. This might not be true in other circumstances. The HQC method provides some compromise between the procedures such as the BIC designed for a true AR situation and the AIC designed for fitting an AR model where the true model may be more general. However, in his recent letter to the present author, E. J. Hannan wrote as follows.

> Quinn and I were a little overenthusiastic in that paper. As I have often pointed out since (see, e.g., Hannan and Deistler [1988, p. 185]), $\ln\ln T$ is a rather meaningless number because it increases so slowly. For T met in practice $(30\leq T\leq 1000)$, $2\ln\ln T$ is around 3 or 4 and its action is just a little stricter than the AIC.

The HQC criterion for ARMA processes is defined by

$$\mathrm{HQC}(k,i) = \ln \tilde{\sigma}^2_{k,i} + 2(k+i)c\frac{\ln\ln T}{T}, \quad c > 1.$$

There are more references about the HQC method in Section 3.9.

3.7 Consistency

In this section we consider the ARMA(p,q) model (3.1). Let K and I be the upper bounds satisfying $p \leq K$ and $q \leq I$, where K and I are known a priori.

Hannan (1980b, 1982) showed the strong consistency of the MBICE and the MHQCE of ARMA processes.

Theorem 3.9. Consider the ARMA(p,q) model (3.1). In addition to the conditions in Section 3.6, we assume that $\{v_t\}$ is a sequence of independent and identically distributed random variables. Then, the MBICE $\hat{p}$ and $\hat{q}$ are strongly consistent. If $\{v_t\}$ is a sequence of uncorrelated random variables; but if $E(\mid v_t \mid^r) < \infty$ for some $r > 4$, then the MBICE $\hat{p}$ and $\hat{q}$ are strongly consistent. □

Theorem 3.10. Consider the ARMA(p,q) model (3.1). In addition to the conditions in Section 3.6, we assume that $\{v_t\}$ is a sequence of independent and identically distributed random variables. Then, the MHQCE's $\hat{p}$ and $\hat{q}$ are strongly consistent. □

Theorem 3.11. Consider the ARMA(p,q) model (3.1). Let $\hat{p}$ and $\hat{q}$ minimize

$$\ln \tilde{\sigma}^2_{k,i} + (k+i)\frac{C(T)}{T},$$

where $C(T)$ satisfies

$$\lim_{T\to\infty} C(T) = \infty, \quad \lim_{T\to\infty} \frac{C(T)}{T} = 0.$$

Then, $\hat{p}$ and $\hat{q}$ are weakly consistent. □

Paulsen (1984) has shown that Theorem 3.11 holds for stationary or nonstationary vector AR processes.

Hannan (1980b) has derived the asymptotic distribution of the MAICE of ARMA processes, which is a generalization of Shibata's asymptotic distribution discussed in Section 3.3.

Theorem 3.12. Consider the ARMA(p,q) model (3.1). Let $\{z_j\}$ be a sequence of independent and identically distributed random variables with

χ_1^2 distribution, and let

$$s_j = \sum_{l=1}^{j}(z_l - 2),$$

$$\pi(k-p, K-k) = P\{s_1 \le 0, \dots, s_{k-p} \le 0\}\{s_1 \ge 0, \dots, s_{K-k} \ge 0\}.$$

Then the MAICE $\hat{p}$ and $\hat{q}$ satisfy the following.

$$\lim_{T\to\infty} P\{\hat{p} = p, \hat{q} = q\} = \pi(i-q, I-i), \quad K = p,\ i \ge q,$$
$$\lim_{T\to\infty} P\{\hat{p} = p, \hat{q} = q\} = \pi(k-p, K-k), \quad k \ge p,\ I = q,$$
$$\lim_{T\to\infty} P\{\hat{p} < p \text{ or } \hat{q} < q\} = 0. \qquad \square$$

As discussed before, when the number of observations T increases, it is reasonable to assume the upper bound K_T of the AR order increases. An, Chen, and Hannan (1982) have shown the strong consistency of the MBICE in this case.

Theorem 3.13. Consider the AR(p) model (3.1), i.e., $q = 0$. In addition to the conditions in Section 3.6, assume that the upper bound K_T of the AR order satisfies

$$K_T = O(\ln T).$$

Then, the MBICE $\hat{p}$ is strongly consistent. $\square$

There are more references about the consistency problem in Section 3.9.

3.8 Some Relations

3.8.1 A Bayesian Interpretation

As mentioned before, Schwarz' BIC does not depend on the prior density function. Thus, Chow (1981) questioned the use of only the first two terms of $S(\bar{\mathbf{y}}, T, i)$ as a model selection criterion. Poskitt and Tremayne (1983, 1987) have asserted the prior probability p_i that the ith model is the true one must be specified in order to employ the posterior to select a model. They have shown that each penalty function statistic can be regarded as a Bayesian criterion with a suitable prior as follows:

$$\mathrm{AIC}_\alpha \quad : \quad p_i \propto \left\{\frac{T}{2\pi}\right\}^{k_i/2} \exp\left(-\frac{1}{2}k_i\alpha\right),$$

$$\mathrm{BIC} \quad : \quad p_i \propto \left\{\frac{1}{2\pi}\right\}^{k_i/2},$$

$$\text{HQC} \quad : \quad p_i \propto \left\{ \frac{T}{2\pi} \right\}^{k_i/2} \exp\left(-\frac{1}{2} k_i c \ln \ln T \right).$$

3.8.2 The BIC and Prediction Errors

Rissanen (1984,1986a, 1986b) has interpreted the BIC using minimization of the accumulated prediction error. Let $\{y_t\}$ be from an AR(p) model, and let $\hat{y}_{t+1}$ be any predictor of y_{t+1}, which is a measurable function of the past observations $y_1, \ldots, y_t$. As discussed in Section 3.1, for any ϵ (> 0), there exists M such that for any $T > M$,

$$E\left\{(\hat{y}_{T+1} - y_{T+1})^2\right\} \geq \sigma^2 \left(1 + \frac{p - \epsilon}{T}\right).$$

It is usual to minimize the nonaccumulated prediction error for obtaining the best linear predictor. We minimize an unbiased estimate of the RHS of the above inequality, which is asymptotically equivalent to the FPE and the AIC. However, Rissanen has proposed to use the accumulated prediction error instead of the nonaccumulated prediction error in statistical modeling. The accumulated prediction error is defined by

$$\frac{1}{T} \sum_{t=0}^{T-1} (\hat{y}_{t+1} - y_{t+1})^2.$$

Rissanen (1986b, p. 58) has explained why he prefers the accumulated prediction error to the nonaccumulated prediction error as follows.

> Applying the sensible reasoning that we should act on the principle that has worked best in the past (indeed, we cannot think of a better principle for statistical inference!), these estimates should clearly be determined by minimization of the summed past prediction errors. Thus, the number of parameters is then estimated so that the accumulated prediction error is minimized.

Rissanen (1986a, p. 1089) has derived the following theorem using information theory.

Theorem 3.14. Let $\{y_t\}$ be the Gaussian ARMA(p, q) process, where the parameter vector $\boldsymbol{\alpha} = (\phi_1, \ldots, \phi_p, \theta_1, \ldots, \theta_q, \sigma^2)^t$ ranges over a compact subset Ω of the $(p + q + 1)$-dimensional Euclidean space with nonempty interior. Let $\hat{y}_{t+1}$ be any predictor of y_{t+1} as a measurable function of $y_1, \ldots, y_t$. Then, for any ϵ (> 0) and for any $\boldsymbol{\alpha} \in \Omega$ except in a set $A_\epsilon(T)$, whose volume shrinks to 0 as T tends to ∞, the following inequality holds:

$$\frac{1}{T} \sum_{t=0}^{T-1} E_{\boldsymbol{\alpha}}\left\{(\hat{y}_{t+1} - y_{t+1})^2\right\} \geq \sigma^2 \left\{1 + (p + q - \epsilon)\frac{\ln T}{T}\right\}. \quad \square$$

The RHS of the last inequality is called Rissanen's lower bound. From Theorem 3.14 we know that minimizing an unbiased estimate of the accumulated prediction error is equivalent to minimizing the BIC. Kavalieris (1989) has presented a polished version of Theorem 3.14. Let $\{y_1, \ldots, y_T\}$ be a T-realization from the AR(∞) model

$$y_t = \sum_{j=1}^{\infty} \phi_j y_{t-j} + v_t,$$

where

$$\phi_\infty(z) = -\sum_{j=0}^{\infty} \phi_j z^j, \quad \phi_0 = -1,$$

is analytical for $\mid z \mid \leq 1$,

$$\sum_{j=0}^{\infty} j \mid \phi_j \mid < \infty,$$

and $\{v_t\}$ is a sequence of independent and identically distributed random variables satisfying

$$E(v_t) = 0, \quad E(v_t^2) = \sigma^2, \quad E(\mid v_t \mid^r) < \infty, \ r = 3, 4, \ldots.$$

For each k, define $\phi_{k,1}, \ldots, \phi_{k,k}$ by the coefficients minimizing

$$E\left\{ y_t - \sum_{j=1}^{k} \phi_{k,j} y_{t-j} \right\}^2,$$

and denote its minimum by σ_k^2. The recursive residuals are defined by

$$\tilde{v}_{k,t+1} = y_{t+1} - \sum_{j=1}^{k} \hat{\phi}_{k,j}(t) y_{t+1-j},$$

where $\hat{\phi}_{k,1}(t), \ldots, \hat{\phi}_{k,k}(t)$ are the OLS estimates of the AR parameters obtained by fitting the AR(k) model to the observations $y_1, \ldots, y_t$ with assumption that $y_s = 0$ for $s \leq 0$. Here (t) indicates that the observations available up to time t are used to compute $\{\hat{\phi}_{k,j}(t) \mid j = 1, \ldots, k\}$. Define an estimate of σ_k^2 by

$$\hat{\sigma}_k^2 = \frac{1}{T} \sum_{s=1}^{T} \left\{ y_s - \sum_{j=1}^{k} \hat{\phi}_{k,j}(T) y_{s-j} \right\}^2,$$

which is the YW estimate obtained by fitting the AR(k) model to $\{y_1, \ldots, y_T\}$. The following theorem shows that minimizing the sum of squares of recursive residuals is asymptotically equivalent to minimizing the BIC.

Theorem 3.15. For any k,

$$\frac{1}{T}\sum_{t=1}^{T}\tilde{v}_{k,t}^2 = \hat{\sigma}_k^2 + (k\sigma_k^2 + c_k)\frac{\ln T}{T}\{1+o(1)\}, \text{ a.s.,}$$

where $0 \le c_k \le (\sigma_k^2 - \sigma^2)c$ for some constant c. □

Rissanen (1986b) has remarked why the BIC based on the accumulated prediction error is consistent, whereas the AIC based on the nonaccumulated prediction error is inconsistent as follows. The criterion based on the accumulated prediction error forces a validation after each observation is received, which leads to a greater penalty on the number of parameters used. Thus, it should produce consistent order estimates.

3.8.3 The AIC and Cross-Validations

Stone (1974) proposed the notion of cross-validation in regression analysis. The residual is the error in predicting the present value from the past observations. However, cross-validation methods use the error appearing when the present value is predicted from the past and the future observations. There have been significant concerns for the use of cross-validation methods in time series analysis. Particularly, Stone (1977) and Stoica, Eykhoff, Janssen, and Söderstöm (1986) have shown the relationship between the AIC and cross-validation. Kavalieris (1989) has refined it as follows.

Let $\{y_1, \ldots, y_T\}$ be a T-realization from the AR(∞) model

$$y_t = \sum_{j=1}^{\infty}\phi_j y_{t-j} + v_t.$$

As in the previous subsection, we assume that

$$\phi_\infty(z) = -\sum_{j=0}^{\infty}\phi_j z^j, \quad \phi_0 = -1,$$

is analytical for $| z | \le 1$,

$$\sum_{j=0}^{\infty} j \mid \phi_j \mid < \infty,$$

and $\{v_t\}$ is a sequence of independent and identically distributed random variables with

$$E(v_t) = 0, \quad E(v_t^2) = \sigma^2, \quad E(\mid v_t \mid^r) < \infty, \ r = 3, 4, \ldots.$$

We follow the notations in the previous subsection. Also, let $\check{\phi}_{k,1}^{[i]}, \ldots, \check{\phi}_{k,k}^{[i]}$

be the estimates minimizing

$$\sum_{t \neq i}^{T} \left\{ y_t - \sum_{j=1}^{k} \phi_{k,j}^{[i]} y_{t-j} \right\}^2,$$

and let

$$\check{v}_{k,t} = y_t - \sum_{j=1}^{k} \check{\phi}_{k,j}^{[t]} y_{t-j}.$$

The cross-validation identification is to choose the order k minimizing

$$\sum_{t=1}^{T} \check{v}_{k,t}^2.$$

Theorem 3.16. Let $K_T < T^\alpha$ and $\alpha < \frac{1}{2}$. Then, uniformly for $k < K_T$,

$$\frac{1}{T} \sum_{t=1}^{T} \check{v}_{k,t}^2 = \hat{\sigma}_k^2 + (k\sigma_k^2 + c_k)\frac{2}{T}\{1 + o(1)\}, \text{ a.s.,}$$

where $0 \leq c_k \leq (\sigma_k^2 - \sigma^2)c$ for some constant c. □

Theorem 3.16 explains why the cross-validation statistic has a tendency toward overfitting. There are more references about the relationship between the AIC and cross-validations in Section 3.9.

3.9 Additional References

- (Section 3.1) About choosing upper bounds of the AR and the MA orders, readers may refer to An, Chen, and Hannan (1982), Hannan and Rissanen (1982), Hannan and Kavalieris (1984b, 1986a), Poskitt (1987), Hannan and Deistler (1988), and the references therein.

- (Section 3.1) The FPE procedure has been used for statistical modeling beyond AR order determination. McClave (1975) utilized Hocking and Leslies' subset regression technique (1967) with the FPE for AR model identification. The FPE procedure was studied for vector AR processes by Akaike (1971), Reinsel (1980, 1983), and Jones (1976), and for more general stochastic processes by Baillie (1979b) and Toyooka (1982). Hsiao (1979) used it for Granger causality tests. Akaike (1969b, 1970b), Gersch and Sharpe (1973), and Jones (1974) used the AR model and the FPE criterion to estimate spectral densities. Their numerical examples have shown that the MFPEE and the

YW estimates of AR coefficients result in reasonable spectral density estimates. However, some examples in disagreement were presented by Marple (1980).

- (Section 3.1) For more details about the asymptotic mean square error of a multi-step ahead predictor, readers may refer to Box and Jenkins (1976, p. 267), Bloomfield (1972), Bhansali (1978), Schmidt (1974), Janacek (1975), Yamamoto (1976, 1981), Baillie (1979a), Davies and Newbold (1980), Reinsel (1980), Shibata (1980), Ledolter and Abraham (1981), Fuller and Hasza (1981), Newton and Pagano (1983), Fotopoulos and Ray (1983), and the references therein.

- (Section 3.2) Readers who are interested in Sanov's theorem may refer to Bahadur and Zabell (1979), Vincze (1982), Deuschel and Strook (1989), and the references therein.

- (Section 3.2) The conditional probability characterization of the Kullback-Leibler information number has been discussed by Vasicek (1980), Csiszár, Cover and Choi (1987), and Choi (1991b).

- (Section 3.2) The AIC was used to select optimal models in many fields of statistics. Akaike (1971-1983), Gersch and Kitagawa (1983), and others utilized it to determine the orders of ARMA processes. Kitagawa (1981) applied the AIC to model fitting for nonstationary time series. Kozin and Nakajima (1980) used the AIC for time-varying AR models. Gabr and Subba Rao (1981) applied it to bilinear time series models. Jones (1974), Sakai (1981), Quinn (1980b, 1988), and Paulsen and Tjøstheim (1985) proposed using the AIC for determining the orders of vector AR processes. It was also used in factor analysis by Akaike (1972b, 1975) and Tong (1975a), in regression analysis by Sawa (1978) and Shibata (1981a, 1984), in the analysis of Markov processes by Tong (1975b), in the analysis of distributed lag model by Tong (1976), in the analysis of covariance by Akaike (1977a), in signal processing analysis by Tong (1975a, 1977) and Findley (1984), and for determining the histogram width by Taylor (1987). Other possible applications have been suggested by Akaike (1973a, 1977a) and Sugiura (1978).

- (Section 3.3) For more details of the asymptotic distribution of the MAICE, refer to Hannan and Deistler (1988, Section 5.6). Sakai (1981), Paulsen and Tjøstheim (1985), and Quinn (1988) derived the asymptotic distribution of the MAICE for vector AR processes.

- (Section 3.4) Some illustrative examples of the CAT procedure were given by Parzen (1979a, 1979b, 1980a) and Parzen and Pagano (1979). The CAT for vector AR processes has been proposed by Parzen (1977) and Parzen and Newton (1980).

- (Section 3.4) For E. J. Hannan's opinion about the AIC, refer to Hannan and Quinn (1979, p. 195) and Hannan (1980b, p. 1072); (1982, p. 411).

- (Section 3.6) For the HQC method, readers may also refer to Heyde and Scott (1973) and Bai, Subramanyam, and Zhao (1988). Quinn (1980b) has generalized the MHQCE to vector AR models and has shown its strong consistency.

- (Section 3.7) For more details about the consistency problem of the penalty function methods, refer to Hannan (1981), Hannan and Deistler (1988, Section 5.4), An and Chen (1986), and the references therein. Pötscher (1990) has shown that if $\hat{r}$ is the estimate of $r = \max(p, q)$ having the first "local" minimum of the BIC under the assumption $k = i$, then $\hat{r} \to r$ a.s. as $T \to \infty$.

- (Section 3.8) Rissanen (1986b) has derived Rissanen's lower bound using coding theory. Kabaila (1987) has shown that under some fairly strong restrictions it can be derived *via* the Cramer-Rao lower bound or the Fisher bound on asymptotic variances for the case of Gaussian AR processes. Also, refer to Hannan, McDougall, and Poskitt (1989).

- (Section 3.8) There have been some recent advances in using cross-validation procedures in time series analysis. Some applications have been considered by Geisser and Eddy (1979), Bessler and Binkley (1980), and Hjorth and Holmqvist (1981). Hurvich and Beltrão (1990) have presented the cross-validated log-likelihood criterion, which can be viewed as a cross-validatory generalization of the AIC. Also, refer to Hurvich and Zeger (1990). Stoica, Eykhoff, Janssen, and Söderström (1986) have presented another cross-validation method, which yields asymptotically the same result as the BIC procedure. Also, refer to Jong (1988).

- (Section 3.8) Tjøstheim and Paulsen (1985) have applied the penalty function identification methods to a particular nonstationary AR process, where the variance of the innovation process depends on time.

4

Innovation Regression Methods

4.1 AR and MA Approximations

A penalty function identification of ARMA processes is to choose the orders minimizing

$$\ln \breve{\sigma}_{k,i}^2 + (k+i)\frac{C(T)}{T}$$

among $k = 0, \ldots, K$ and $i = 0, \ldots, I$. Here $\breve{\sigma}_{k,i}^2$ is an estimate of the innovation variance obtained by fitting the ARMA(k, i) model to the observations and K and I are determined a priori as upper bounds of the orders. Because there are $(K+1) \times (I+1)$ possible ARMA models to be estimated, it is computationally onerous to apply ML estimation methods. Even though many algorithms have been presented to obtain the exact ML estimates as mentioned in Chapter 1, there are still many problems in applying them to all the possible ARMA models. Especially if the MA part exists, then the ML estimates are not always on the stationary and invertible region. They are very sensitive to the quality of starting values for the algorithms. Also, the algorithms do not always produce the ML estimates, for they often do not converge, particularly with improper initial values. Hannan, Dunsmuir, and Deistler (1980) have shown that if $k \geq p$ and $i \geq q$, then $\tilde{\psi}_{k,i}(z) = \tilde{\phi}_k^{-1}(z)\tilde{\theta}_i(z)$, where $\tilde{\phi}_k(z)$ and $\tilde{\theta}_i(z)$ are the ML estimates of $\phi(z)$ and $\theta(z)$, respectively, converges to $\psi(z) = \phi^{-1}(z)\theta(z)$ uniformly on the closed unit disc. However, if $k > p$ and $i > q$, then $\tilde{\phi}_k(z)$ and $\tilde{\theta}_i(z)$ do not converge to $\phi(z)$ and $\theta(z)$ in any reasonable sense. Because the likelihood is constant along the "line" $\tilde{\psi} = \psi$, the sample point will search up and down that line as T increases. Therefore, the ML method fails to provide consistent estimates of the parameters if $k > p$ and $i > q$. There are more references about this inconsistency of the ML estimates in Section 4.5.

Recently, some linear estimation procedures have been suggested for ARMA modeling, which determine the orders as well as estimate the parameters. Their identification methods are using the penalty functions discussed in Chapter 3. Because the optimization problems in these methods are quadratic, they are computationally much cheaper than the exact ML method. If we use one of them to procreate initial estimates for ML algorithms, then we can keep the ML algorithms not only from diverging but also from converging to nonstationary and/or noninvertible regions. However, there should be some discriminations among the linear estimation procedures based on their stabilities. The linear estimation methods have been also applied to general transfer function models and the ARMAX model. Interested readers may consult the references in Section 4.5.

The linear estimation methods consist of two steps. The first step is to fit an AR model with high order, which is sometimes called a long autoregression, to the observations. The second step is to apply the OLS method to the observations and the estimated innovations of the long AR model. In this section we consider the first step in detail.

Because the ARMA(p, q) process is assumed to be invertible, it can be approximated by an AR(n) model for large n. This method was originated by Durbin (1959), who fitted a long AR model to the observations in order to estimate MA processes. For more details about the long AR method, readers may consult the references in Section 4.5.

As discussed in Chapter 1, the YW estimates are more biased than the OLS and Burg's estimates. Thus, for a long autoregression, we would rather use either OLS estimates or Burg's estimates than the YW estimates. Wahlberg (1989a, 1989b) has shown that efficient estimates of the ARMA parameters can be obtained through a long AR(n) model only if n tends to ∞ as T does. Thus, we can show that efficient estimates of the ARMA model using the sample autocovariances can be obtained only if the number of the sample autocovariance terms tends to ∞ as T does. (See also Porat [1986].) To study asymptotic behaviors of the long AR model estimates, Hannan and his colleagues as well as Saikkonen (1986) have assumed

$$n = O\{(\ln T)^{\alpha}\}, \quad \alpha > 1,$$

as discussed in Chapter 3, whereas Berk (1974), Bhansali (1978), and Wahlberg (1989a) have restricted the growth rate of n to

$$\lim_{T \to \infty} \frac{n^3}{T} = 0.$$

Under the assumption, Saikkonen has shown that the OLS estimates of the ARMA parameters through a long AR model are strongly consistent. Durbin (1959) has shown the asymptotic efficiency of the OLS estimates of pure MA processes, and Saikkonen has shown that they are inefficient in case of mixed ARMA processes.

Saikkonen (1986) has proposed to apply an OLS method to a long MA model for estimating the parameters of ARMA models. The long MA model estimates are strongly consistent, and are efficient for pure AR processes but inefficient for ARMA processes. The asymptotic efficiency of the long MA model estimates is based on the assumption that the order of the MA model tends to ∞ as T does.

4.2 Hannan and Rissanen's Method

Hannan and Rissanen (1982), referred to as HR-82, have proposed a three-stage ARMA modeling method by fitting the current observation to past observations and estimated values of past innovations. It will be called the HR method. In this section, we assume that the observations $\{y_1, \ldots, y_T\}$ are from the Gaussian ARMA(p, q) model,

$$\phi(B)y_t = \theta(B)v_t, \tag{4.1}$$

where $\phi(B) = -\phi_0 - \phi_1 B - \cdots - \phi_p B^p$, $\theta(B) = -\theta_0 - \theta_1 B - \cdots - \theta_q B^q$, $\phi_0 = \theta_0 = -1$, $\phi_p \neq 0$, $\theta_q \neq 0$, and the white noise process $\{v_t\}$ is Gaussian with variances σ^2 (> 0). We assume that the model is stationary, invertible, and coprimal. As Hannan has emphasized in several literatures, it is hardly convincing that the ARMA model in (4.1) is a true one. We would rather consider it as representing a more delicate natural system. However, in this section we assume that the model (4.1) is a precise one.

4.2.1 A Three-Stage Procedure

The HR method of ARMA modeling is a three-stage method as follows.

1. The innovation $\check{v}_t$ is calculated by fitting a long AR model. Let N_T be an upper bound of the AR orders such that N_T increases as T does but it does not increase too quickly. It is sufficient to assume

$$N_T < (\ln T)^a, \quad 0 < a < \infty.$$

For n $(\leq N_T)$, fit the AR(n) model

$$y_t = \sum_{l=1}^{n} \beta_l y_{t-l} + v_t$$

to the observations $\{y_1, \ldots, y_T\}$. Let $\check{\beta}_1, \ldots, \check{\beta}_n$ be the YW estimates of $\beta_1, \ldots, \beta_n$. Then, calculate the estimated innovations

$$\check{v}_t = -\sum_{l=0}^{n} \check{\beta}_l y_{t-l}, \quad t = n+1, \ldots, T,$$

where $\check{\beta}_0 = -1$. It is recommended to choose the AR order n using the BIC or the AIC. Due to its overparameterization tendency, the AIC may be more practical.

2. Let K and I be sufficiently large so that they are greater than the true orders p and q, respectively. For each $(k, i) \in \{(k, i) \mid k =$

$0, \ldots, K,\ i = 0, \ldots, I\}$, calculate the OLS estimates $\tilde{\phi}_{k,1}, \ldots, \tilde{\phi}_{k,k}$, $\tilde{\theta}_{i,1}, \ldots, \tilde{\theta}_{i,i}$ minimizing

$$\frac{1}{T} \sum_{t=t_0+1}^{T} \left(y_t - \sum_{j=1}^{k} \phi_{k,j} y_{t-j} + \sum_{l=1}^{i} \theta_{i,l} \check{v}_{t-l} \right)^2,$$

where $t_0 = \max(n+k, n+i)$. Denote its minimum by $\tilde{\sigma}^2_{k,i}$. Choose $\tilde{p}$ and $\tilde{q}$ minimizing

$$\mathrm{BIC}(k,i) = \ln \tilde{\sigma}^2_{k,i} + (k+i)\frac{\ln T}{T}$$

among $k = 0, \ldots, K$ and $i = 0, \ldots, I$. The estimated innovations in this stage are defined by

$$\tilde{v}_t = y_t = 0, \quad t \leq 0,$$

$$\tilde{v}_t = y_t - \sum_{j=1}^{\tilde{p}} \tilde{\phi}_{\tilde{p},j} y_{t-j} + \sum_{l=1}^{\tilde{q}} \tilde{\theta}_{\tilde{q},l} \tilde{v}_{t-l}, \quad t = 1, \ldots, T.$$

3. Using the estimates of the second stage as initial values, we apply ML techniques. One of them is as follows. Calculate $\{x_t\}$ and $\{z_t\}$ by

$$x_t = z_t = 0, \quad t \leq 0,$$

$$x_t = -\sum_{l=1}^{\tilde{q}} \tilde{\theta}_{\tilde{q},l} x_{t-l} + y_t, \quad t = 1, \ldots, T,$$

$$z_t = -\sum_{l=1}^{\tilde{q}} \tilde{\theta}_{\tilde{q},l} z_{t-l} + \tilde{v}_t, \quad t = 1, \ldots, T.$$

Regress $\tilde{v}_t + x_t - z_t$ on

$$x_{t-1}, \ldots, x_{t-\tilde{p}}, -z_{t-1}, \ldots, -z_{t-\tilde{q}}$$

for $t = t_1 + 1, \ldots, T$, where $t_1 = \max(\tilde{p}, \tilde{q})$ in order to obtain the refined estimates $\hat{\phi}_{\tilde{p},1}, \ldots, \hat{\phi}_{\tilde{p},\tilde{p}}, \hat{\theta}_{\tilde{q},1}, \ldots, \hat{\theta}_{\tilde{q},\tilde{q}}$. The corresponding estimate of the white noise variance is

$$\hat{\sigma}^2_{\tilde{p},\tilde{q}} = \frac{1}{T} \sum_{t=t_1+1}^{T} \left\{ y_t - \sum_{j=1}^{\tilde{p}} \hat{\phi}_{\tilde{p},j} y_{t-j} + \sum_{l=1}^{\tilde{q}} \hat{\theta}_{\tilde{q},l} \hat{v}_{t-l} \right\}^2,$$

where $\hat{v}_t$ is the reestimated residuals.

The third stage might be iterated by replacing $\tilde{\phi}_{\tilde{p},1}, \ldots, \tilde{\phi}_{\tilde{p},\tilde{p}}$ and $\tilde{\theta}_{\tilde{q},1}, \ldots, \tilde{\theta}_{\tilde{q},\tilde{q}}$ with $\hat{\phi}_{\tilde{p},1}, \ldots, \hat{\phi}_{\tilde{p},\tilde{p}}$ and $\hat{\theta}_{\tilde{q},1}, \ldots, \hat{\theta}_{\tilde{q},\tilde{q}}$ until $\hat{\sigma}^2_{\tilde{p},\tilde{q}}$ stabilizes. The three-stage method is essentially due to Durbin (1960b), who, however, gave no rule for determining the long AR order n and took p and q given. There have been other practices of Durbin's method. Interested readers may consult the references in Section 4.5.

4.2.2 Block Toeplitz Matrices

In the second stage of the HR method, it is necessary to solve a system of block Toeplitz equations. Because the algebraic structure of Toeplitz matrices is very important in linear prediction and the related fields, we discuss it separately in this subsection.

Block Toeplitz matrices appear as correlation matrices of multivariate stationary processes. A $p \times q$ block matrix of basic dimension $d \times e$ is defined by a $pd \times qe$ matrix composed of $d \times e$ submatrices. Throughout this subsection each matrix is assumed to be partitioned by $d \times d$ submatrices. The basic dimension will not be described, unless there is any possibility of confusion about it.

Consider a sequence of nested block Toeplitz matrices $\{\boldsymbol{C}_k : k = 1, 2, \ldots\}$, where $\boldsymbol{C}_k$ is a $k \times k$ block matrix defined by

$$\boldsymbol{C}_k = \begin{bmatrix} C_0 & C_{-1} & \cdots & C_{-k+1} \\ C_1 & C_0 & \cdots & C_{-k+2} \\ \vdots & \vdots & & \vdots \\ C_{k-1} & C_{k-2} & \cdots & C_0 \end{bmatrix}.$$

Assume that $\boldsymbol{C}_k$ is nonsingular for $k = 1, 2, \ldots$. The purpose of this subsection is to represent the inverse of $\boldsymbol{C}_k$ as a product of an upper block triangular, a block diagonal, and a lower block triangular matrices and to present a recursive algorithm for the decomposition. The blocks of the component matrices are the solutions of the forward and the backward EYW equations of associated vector ARMA processes, and the recursive algorithm is a generalization of the Levinson-Durbin one.

If $\boldsymbol{A}$ is a $p \times q$ block matrix whose (i, j) block is $A_{i,j}$, then the block transposed matrix $\boldsymbol{A}^T$ is defined as a $q \times p$ block matrix whose (i, j) block is $A_{j,i}$. Also, the elementwise transposed matrix of $\boldsymbol{A}$ is denoted by $\boldsymbol{A}^t$. The block exchange matrix $\boldsymbol{E}_n$ is defined as the $n \times n$ block matrix whose (i, j) block is the identity matrix I_d if $i + j = n + 1$, and O otherwise. Also, let $\tilde{\boldsymbol{A}} = \boldsymbol{A}\boldsymbol{E}_q$.

For $k = 1, 2, \ldots$ and $j = 1, \ldots, k$, let $L_{k,j}$, $U_{k,j}$, $M_{k,j}$, and $N_{k,j}$ be $d \times d$ matrices satisfying the simultaneous equations:

$$\begin{aligned} C_j &= L_{k,1}C_{j-1} + L_{k,2}C_{j-2} + \cdots + L_{k,k}C_{j-k}, \quad j = 1, \ldots, k, \\ C_j &= C_{j+1}U_{k,1} + C_{j+2}U_{k,2} + \cdots + C_{j+k}U_{k,k}, \quad j = -1, \ldots, -k, \\ C_j &= M_{k,1}C_{j+1} + M_{k,2}C_{j+2} + \cdots + M_{k,k}C_{j+k}, \quad j = -1, \ldots, -k, \\ C_j &= C_{j-1}N_{k,1} + C_{j-2}N_{k,2} + \cdots + C_{j-k}N_{k,k}, \quad j = 1, \ldots, k. \end{aligned}$$

The nonsingularity of $\boldsymbol{C}_k$ implies the solutions exist uniquely. If $\{C_k \mid k = 0, 1, \ldots\}$ is the ACVF of a vector ARMA process, then the first and the fourth equations are the forward EYW equations and the second and the third are the backward EYW equations. Let

$$L_{k,0} = U_{k,0} = M_{k,0} = N_{k,0} = -I_d$$

for $k = 0, 1, \ldots$. Define lower block triangular matrices $\boldsymbol{L}_k$, $\boldsymbol{M}_k$, upper block triangular matrices $\boldsymbol{U}_k$, $\boldsymbol{N}_k$, and block diagonal matrices $\triangle_k$ and ∇_k by

$$\boldsymbol{L}_k = \begin{bmatrix} L_{0,0} & O & O & \cdots & O \\ L_{1,1} & L_{1,0} & O & \cdots & O \\ L_{2,2} & L_{2,1} & L_{2,0} & \cdots & O \\ \vdots & \vdots & \vdots & & \vdots \\ L_{k-1,k-1} & L_{k-1,k-2} & L_{k-1,k-3} & \cdots & L_{k-1,0} \end{bmatrix},$$

$$\boldsymbol{M}_k = \begin{bmatrix} M_{0,0} & O & O & \cdots & O \\ M_{1,1} & M_{1,0} & O & \cdots & O \\ M_{2,2} & M_{2,1} & M_{2,0} & \cdots & O \\ \vdots & \vdots & \vdots & & \vdots \\ M_{k-1,k-1} & M_{k-1,k-2} & M_{k-1,k-3} & \cdots & M_{k-1,0} \end{bmatrix},$$

$$\boldsymbol{U}_k = \begin{bmatrix} U_{0,0} & U_{1,1} & U_{2,2} & \cdots & U_{k-1,k-1} \\ O & U_{1,0} & U_{2,1} & \cdots & U_{k-1,k-2} \\ O & O & U_{2,0} & \cdots & U_{k-1,k-3} \\ O & O & O & \cdots & U_{k-1,0} \end{bmatrix},$$

$$\boldsymbol{N}_k = \begin{bmatrix} N_{0,0} & N_{1,1} & N_{2,2} & \cdots & N_{k-1,k-1} \\ O & N_{1,0} & N_{2,1} & \cdots & N_{k-1,k-2} \\ O & O & N_{2,0} & \cdots & N_{k-1,k-3} \\ O & O & O & \cdots & N_{k-1,0} \end{bmatrix},$$

$$\triangle_k = \begin{bmatrix} \Lambda_0^{-1} & O & O & \cdots & O \\ O & \Lambda_1^{-1} & O & \cdots & O \\ O & O & \Lambda_2^{-1} & \cdots & O \\ \vdots & \vdots & \vdots & & \vdots \\ O & O & O & \cdots & \Lambda_{k-1}^{-1} \end{bmatrix},$$

$$\nabla_k = \begin{bmatrix} V_0^{-1} & O & O & \cdots & O \\ O & V_1^{-1} & O & \cdots & O \\ O & O & V_2^{-1} & \cdots & O \\ \vdots & \vdots & \vdots & & \vdots \\ O & O & O & \cdots & V_{k-1}^{-1} \end{bmatrix},$$

where

$$\Lambda_0 = V_0 = C_0,$$

$$\Lambda_i = -(L_{i,0}C_0 + L_{i,1}C_{-1} + \cdots + L_{i,i}C_{-i}), \quad i = 1, \ldots, k-1,$$
$$V_i = -(M_{i,0}C_0 + M_{i,1}C_1 + \cdots + M_{i,i}C_i), \quad i = 1, \ldots, k-1.$$

It should be noted that $\boldsymbol{L}_k$, $\boldsymbol{M}_k$, $\boldsymbol{U}_k$, $\boldsymbol{N}_k$, $\triangle_k$, and $\bigtriangledown_k$ are nested by $\boldsymbol{L}_{k+1}$, $\boldsymbol{M}_{k+1}$, $\boldsymbol{U}_{k+1}$, $\boldsymbol{N}_{k+1}$, $\triangle_{k+1}$, and $\bigtriangledown_{k+1}$, respectively.

Theorem 4.1. For $k = 1, 2, \ldots$, the inverse of a block Toeplitz matrix can be uniquely factored into a product of an upper block triangular, a block diagonal, and a lower block triangular matrices as follows:

$$\boldsymbol{C}_k^{-1} = \boldsymbol{U}_k \triangle_k \boldsymbol{L}_k,$$

$$(\boldsymbol{C}_k^T)^{-1} = \boldsymbol{N}_k \bigtriangledown_k \boldsymbol{M}_k. \quad \square$$

To derive the recursive algorithm for the decompositions in Theorem 4.1, we define some block matrices and present their properties. For $k = 1, 2, \ldots$, let

$$L_k = (L_{k,1}, L_{k,2}, \ldots, L_{k,k})^T,$$
$$U_k = (U_{k,1}, U_{k,2}, \ldots, U_{k,k})^T,$$
$$M_k = (M_{k,1}, M_{k,2}, \ldots, M_{k,k})^T,$$
$$N_k = (N_{k,1}, N_{k,2}, \ldots, N_{k,k})^T,$$
$$S_k = (C_1, C_2, \ldots, C_k)^T,$$
$$T_k = (C_{-k}, C_{1-k}, \ldots, C_{-1})^T,$$
$$G_k = C_{k+1} - \tilde{S}_k^T \boldsymbol{C}_k^{-1} S_k,$$
$$H_k = C_{-k-1} - \tilde{T}_k^T \boldsymbol{C}_k^{-1} T_k.$$

Then

$$\tilde{L}_k^T = \tilde{S}_k^T \boldsymbol{C}_k^{-1},$$
$$\tilde{U}_k = \boldsymbol{C}_k^{-1} T_k,$$
$$M_k^T = \tilde{T}_k^T \boldsymbol{C}_k^{-1},$$
$$N_k = \boldsymbol{C}_k^{-1} S_k,$$
$$G_k = C_{k+1} - \tilde{L}_k^T S_k = C_{k+1} - \tilde{S}_k^T N_k,$$
$$H_k = C_{-k-1} - \tilde{T}_k^T \tilde{U}_k = C_{-k-1} - M_k^T T_k,$$
$$\Lambda_k = C_0 - \tilde{L}_k^T T_k = C_0 - \tilde{S}_k^T \tilde{U}_k,$$
$$V_k = C_0 - M_k^T S_k = C_0 - \tilde{T}_k^T N_k.$$

Using the definitions and the properties, we can derive the following recursive algorithm.

Algorithm 4.1. *An Algorithm for Block Toeplitz Matrices*
As the initialization part, let

$$\begin{aligned} L_{0,0} &= U_{0,0} = M_{0,0} = N_{0,0} = -I_d, \\ G_0 &= C_1, \\ H_0 &= C_{-1}, \\ \Lambda_0 &= V_0 = C_0. \end{aligned}$$

For $k = 0, 1, \ldots$, calculate

$$\begin{aligned} L_{k+1} &= \begin{pmatrix} L_k - L_{k+1,k+1}\tilde{M}_k \\ G_k V_k^{-1} \end{pmatrix}, \\ U_{k+1} &= \begin{pmatrix} U_k - \tilde{N}_k U_{k+1,k+1} \\ V_k^{-1} H_k \end{pmatrix}, \\ M_{k+1} &= \begin{pmatrix} M_k - M_{k+1,k+1}\tilde{L}_k \\ H_k \Lambda_k^{-1} \end{pmatrix}, \\ N_{k+1} &= \begin{pmatrix} N_k - \tilde{U}_k N_{k+1,k+1} \\ \Lambda_k^{-1} G_k \end{pmatrix}, \\ G_{k+1} &= C_{k+2} - \tilde{L}_{k+1}^T S_{k+1}, \\ H_{k+1} &= C_{-k-2} - M_{k+1}^T T_{k+1}, \\ \Lambda_{k+1} &= (I_d - L_{k+1,k+1} M_{k+1,k+1}) \Lambda_k, \\ V_{k+1} &= V_k (I_d - U_{k+1,k+1} N_{k+1,k+1}). \quad \square \end{aligned}$$

The block Toeplitz decompositions in Theorem 4.1 and the corresponding Algorithm 4.1 have many applications.

Watson (1973) presented a recursive algorithm to invert nested block symmetric Toeplitz matrices. Theorem 4.1 and Algorithm 4.1 can be utilized to invert nested Toeplitz matrices recursively as follows:

$$\boldsymbol{C}_{k+1}^{-1} = \begin{pmatrix} \boldsymbol{C}_k^{-1} + \tilde{U}_k \Lambda_k^{-1} \tilde{L}_k^T & -\tilde{U}_k \Lambda_k^{-1} \\ -\Lambda_k^{-1} \tilde{L}_k^T & \Lambda_k^{-1} \end{pmatrix}.$$

Consider a block Toeplitz system of simultaneous equations

$$\boldsymbol{C}_k X_k = \boldsymbol{B}_k,$$

where $\boldsymbol{B}_k = (B_1, B_2, \ldots, B_k)^T$ is a known $k \times 1$ block matrix and $X_k = (X_{k,1}, X_{k,2}, \ldots, X_{k,k})^T$ is an unknown $k \times 1$ block matrix. Using Theorem 4.1 and Algorithm 4.1 we can easily derive the following:

$$X_{k+1,k+1} = \Lambda_k^{-1}(B_{k+1} - \sum_{j=1}^{k} C_{k+1-j} X_{k,j}),$$

$$X_{k+1,j} = X_{k,j} - U_{k,k+1-j}X_{k+1,k+1}, \quad j = 1, 2, \ldots, k.$$

Because Theorem 4.1 and Algorithm 4.1 are based on the vector EYW solutions, it is natural to link them with vector ARMA processes. If the ACVF of a stationary d-variate ARMA process is denoted by $\{\Sigma_k \mid k = 0, 1, \ldots\}$, then the forward EYW equations are

$$\Sigma_k = A_1\Sigma_{k-1} + A_2\Sigma_{k-2} + A_p\Sigma_{k-p}, \quad k = q+1, q+2, \ldots.$$

If $C_k = \Sigma_{q+k}$, then $A_k = L_{p,k}$ for $k = 1, \ldots, p$. In this case, Algorithm 4.1 is reduced to a refined version of the Whittle algorithm (1963).

We can utilize Theorem 4.1 and Algorithm 4.1 in calculating the determinant of a block Toeplitz matrix as follows:

$$\det(\boldsymbol{C}_k) = \prod_{i=0}^{k-1} \det(\Lambda_i) \prod_{i=0}^{k-1} \det(V_i).$$

This equality will be useful for multivariate model selection through the penalty function identification methods.

Theorem 4.1 and Algorithm 4.1 can be applied to decomposition of a $k \times k$ block Hankel matrix $\boldsymbol{H}_k$ because $\boldsymbol{H}_k\boldsymbol{E}_k$ and $\boldsymbol{E}_k\boldsymbol{H}_k$ are $k \times k$ block Toeplitz matrices.

If the block Toeplitz matrix $\boldsymbol{C}_k$ is elementwise symmetric, i.e., $C_{-j} = C_j^t$ for $j = 0, \ldots, k-1$, then, for each (i, j),

$$U_{i,j} = L_{i,j}^t, \quad N_{i,j} = M_{i,j}^t, \quad H_j = G_j^t,$$

and Λ_j and V_j are symmetric. Thus, Algorithm 4.1 is simplified as follows.

Algorithm 4.2. *An Algorithm for Symmetric Block Toeplitz Matrices*
As the initialization part, let

$$L_{0,0} = M_{0,0} = -I_d,$$
$$G_0 = C_1,$$
$$\Lambda_0 = V_0 = C_0.$$

For $k = 0, 1, \ldots,$ calculate

$$L_{k+1} = \begin{pmatrix} L_k - L_{k+1,k+1}\tilde{M}_k \\ G_k V_k^{-1} \end{pmatrix},$$

$$M_{k+1} = \begin{pmatrix} M_k - M_{k+1,k+1}\tilde{L}_k \\ G_k^t \Lambda_k^{-1} \end{pmatrix},$$

$$G_{k+1} = C_{k+2} - \tilde{L}_{k+1}^T S_{k+1},$$
$$\Lambda_{k+1} = (I_d - L_{k+1,k+1}M_{k+1,k+1})\Lambda_k,$$
$$V_{k+1} = (I_d - M_{k+1,k+1}L_{k+1,k+1})V_k. \quad \square$$

4.2.3 A Modification of the Whittle Algorithm

Hereafter in this section the estimates of the three stages of the HR method will be distinguished by accents. As an example, the first stage, the second stage, and the third stage estimates of μ are denoted by $\check{\mu}, \tilde{\mu}$, and $\hat{\mu}$, respectively. Also, $O(\cdot)$ and $o(\cdot)$ will indicate order relations hold a.s., whereas $O_p(\cdot)$ and $o_p(\cdot)$ will be used when the order relations hold only in probability.

The problem in the second stage of the HR method is to obtain

$$\tilde{\sigma}^2_{k,i} = \inf \frac{1}{T} \sum_{t=t_0+1}^{T} \left\{ y_t - \sum_{j=1}^{k} \phi_{k,j} y_{t-j} + \sum_{l=1}^{i} \theta_{i,l} \check{v}_{t-l} \right\}^2$$

for every pair (k, i) satisfying $k \leq K$ and $i \leq I$. Because the normal equations of the minimization are linear, it is computationally much cheaper than that involved in determining the ML estimates. In some cases, e.g., in automatic control problems, the number of observations may be large, and then the chosen orders will be quite large too. Thus, a computationally cheap algorithm is necessary, which takes the linearity of the problem into account.

HR-82 have recommended the consideration of the problem under the assumption $k = i$. In many situations this suggestion is adequate, at least as a first investigation. As discussed in Section 1.4, if $k = p$ and $i \geq q$ or if $k \geq p$ and $i = q$, then the matrix $\Sigma(k, i)$ is nonsingular. Thus, when $k = i = \max(p, q)$, the ML estimates and the EYW estimates are consistent. Moreover, an unconstrained state space representation of an ARMA model leads to $p = q$. (See, e.g., Aoki [1990, Chapter 4].) After the order $\tilde{h} = \max(\tilde{p}, \tilde{q})$ is chosen under the assumption, we can easily determine $\tilde{p}$ and $\tilde{q}$, if so desired, among either $\tilde{p} = \tilde{h}$ and $\tilde{q} \leq \tilde{h}$ or $\tilde{p} \leq \tilde{h}$ and $\tilde{q} = \tilde{h}$ through various methods including the penalty function identification methods. If the minimum is evaluated over $k = i = 0, \ldots, K(= I)$, then it is necessary only to analyze

$$\min(K, I) + 2\max(p, q) - \min(p, q) + 1$$

pairs of orders instead of $(K + 1) \times (I + 1)$ pairs. Hannan and Kavalieris (1984a, p. 274) have mentioned that this assumption may be also applied to the third stage.

To calculate $\tilde{\sigma}^2_{k,k}$ for $k = 0, \ldots, K$ under the assumption $k = i$, HR-82 have offered to use a modification of the Whittle algorithm, which is a multivariate version of Algorithm 1.2. Here it will be explained using a slightly different way from theirs. Define the sample cross-covariances of $\{y_t\}$ and $\{\check{v}_t\}$ by

$$c_{yy}(j) = c_{yy}(-j) = \frac{1}{T} \sum_{t=1}^{T-j} y_t y_{t+j} \quad j = 0, 1, \ldots,$$

$$\check{c}_{vy}(j) = \check{c}_{yv}(-j) = \frac{1}{T}\sum_{t=1}^{T-j} \check{v}_t y_{t+j}, \quad j = 0, 1, \ldots,$$

$$\check{c}_{yv}(j) = \check{c}_{vy}(-j) = \frac{1}{T}\sum_{t=1}^{T-j} y_t \check{v}_{t+j}, \quad j = 0, 1, \ldots,$$

$$\check{c}_{vv}(j) = \check{c}_{vv}(-j) = \frac{1}{T}\sum_{t=1}^{T-j} \check{v}_t \check{v}_{t+j}, \quad j = 0, 1, \ldots.$$

For $j = 0, \pm 1, \pm 2, \ldots$, $k = 1, 2, \ldots$, $l = 1, \ldots, k$, let

$$C_j = \begin{pmatrix} \check{c}_{vv}(j) & \check{c}_{vy}(j) \\ \check{c}_{yv}(j) & c_{yy}(j) \end{pmatrix}, \quad \boldsymbol{\alpha}_{k,l} = \begin{pmatrix} -\theta_{k,l} \\ \phi_{k,l} \end{pmatrix},$$

Then, $C_{-j} = C_j^t$, $j = 1, 2, \ldots$. Hereafter we will use the same notations as in the previous subsection. If the end terms of the cross-covariances are neglected, then the normal equations for minimizing

$$\frac{1}{T}\sum_{t=t_0+1}^{T} \left\{ y_t - \sum_{j=1}^{k} \phi_{k,j} y_{t-j} + \sum_{l=1}^{k} \theta_{k,l} \check{v}_{t-l} \right\}^2,$$

where $t_0 = n + k$, become

$$\boldsymbol{C}_k \tilde{\boldsymbol{\alpha}}_k = S_k \boldsymbol{i},$$

where

$$\begin{aligned} \boldsymbol{i} &= (0, 1)^t, \\ \tilde{\boldsymbol{\alpha}}_k &= (\tilde{\boldsymbol{\alpha}}_{k,1}, \ldots, \tilde{\boldsymbol{\alpha}}_{k,k})^T, \\ S_k &= (C_1, \ldots, C_k)^T, \end{aligned}$$

and $\boldsymbol{C}_k$ is the $k \times k$ block Toeplitz matrix whose (r, s) block is C_{r-s}. It should be noted that $\tilde{\boldsymbol{\alpha}}_k$ and $\tilde{\boldsymbol{\alpha}}_{k,j}$ are the estimates of $\boldsymbol{\alpha}_k$ and $\boldsymbol{\alpha}_{k,j}$ in the second stage and not the exchanged vectors. The OLS estimate of the coefficient vector is

$$\tilde{\boldsymbol{\alpha}}_k = \boldsymbol{C}_k^{-1} S_k \boldsymbol{i} = N_k \boldsymbol{i},$$

where the last equality holds by the definition of N_k in Subsection 4.2.2. Thus,

$$\tilde{\boldsymbol{\alpha}}_{k,j} = N_{k,j} \boldsymbol{i}, \quad j = 1, \ldots, k.$$

The mean square error becomes

$$\tilde{\sigma}_{k,k}^2 \quad = \quad \sigma(0) - \sum_{j=1}^{k}\sum_{l=1}^{k} \tilde{\boldsymbol{\alpha}}_j^t C_{j-l} \tilde{\boldsymbol{\alpha}}_l$$

$$
\begin{aligned}
&= \sigma(0) - \boldsymbol{i}^t \left(\sum_{j=1}^{k} \sum_{l=1}^{k} \boldsymbol{N}_{k,j}^t \boldsymbol{C}_{j-l} \boldsymbol{N}_{k,l} \right) \boldsymbol{i} \\
&= \sigma(0) - \boldsymbol{i}^t \left(\sum_{j=1}^{k} \boldsymbol{N}_{k,j}^t \boldsymbol{C}_j \right) \boldsymbol{i} \\
&= \sigma(0) - \boldsymbol{i}^t \left(\sum_{j=1}^{k} \boldsymbol{M}_{k,j} \boldsymbol{C}_j \right) \boldsymbol{i},
\end{aligned}
$$

where the last equality holds because $\boldsymbol{C}_k$ is symmetric. We can calculate $\{\tilde{\sigma}_{k,k}^2 \mid k = 0, 1, \ldots\}$ using Algorithm 4.2.

Choi (1991h) has derived a modified procedure of the second stage, which is computationally cheaper than the modified Whittle algorithm.

4.2.4 Some Modifications

As pointed out in a correction by Hannan and Rissanen (1983, p. 303), the second stage of the HR method leads to overestimation of the orders p and q. Hannan and Kavalieris (1984a), referred to as HK-84, called it a bias problem, and suggested some modifications for correcting the bias. Now we examine why the bias occurs. Put

$$
u(B) = -\sum_{l=0}^{q} \theta_l B^l, \quad \theta_0 = -1,
$$

$$
u_t = u(B) v_t,
$$

$$
\omega_j^2 = \min_{\alpha_1, \ldots, \alpha_j} E(u_t - \alpha_1 u_{t-1} - \cdots - \alpha_j u_{t-j})^2, \quad j = 1, 2, \ldots.
$$

For $k \geq p$ and $i \geq q$, let $m = \max(k-p, i-q)$. If either the AIC or the BIC is used to choose the order n of the long AR model in the first stage, then

$$
\tilde{\sigma}_{k,i}^2 = \frac{1}{T} \sum_{t=1}^{T} v_t^2 + \sigma^2 \frac{1}{2 \ln \zeta_0} \frac{\ln T}{T} (\omega_m^2 - 1) + o_p\left(\frac{\ln T}{T} \right),
$$

where ζ_0 is the modulus of a root of $u(z) = 0$ nearest the unit circle. Put

$$
Q_T = \frac{\ln \ln T}{T}.
$$

It is known (see, e.g., Hannan and Deistler [1988, p. 166]) that

$$
\frac{1}{T} \sum_{t=1}^{T} v_t^2 = \sigma^2 + O(Q_T).
$$

Thus, if $k \geq r \geq p$ and $i \geq s \geq q$ and $l = \max(r-p, s-q)$, then

$$\begin{aligned}
&\mathrm{BIC}(k,i) - \mathrm{BIC}(r,s) \\
&= \ln\left(1 + \frac{\tilde{\sigma}^2_{k,i} - \tilde{\sigma}^2_{r,s}}{\tilde{\sigma}^2_{r,s}}\right) + (k+i-r-s)\frac{\ln T}{T} + o_p\left(\frac{\ln T}{T}\right) \\
&= \frac{\tilde{\sigma}^2_{k,i} - \tilde{\sigma}^2_{r,s}}{\tilde{\sigma}^2_{r,s}} + (k+i-r-s)\frac{\ln T}{T} + o_p\left(\frac{\ln T}{T}\right) \\
&= \frac{1}{\sum_{t=1}^T v_t^2/T}\sigma^2 \frac{1}{2\ln\zeta_0}\frac{\ln T}{T}(\omega_m^2 - \omega_l^2) \\
&\quad +(k+i-r-s)\frac{\ln T}{T} + o_p\left(\frac{\ln T}{T}\right) \\
&= \left\{\frac{1}{2\ln\zeta_0}(\omega_m^2 - \omega_l^2) + (k+i-r-s)\right\}\frac{\ln T}{T} + o_p\left(\frac{\ln T}{T}\right).
\end{aligned}$$

It implies the order estimates $\tilde{p}$ and $\tilde{q}$ in the second stage converge to $p+m_0$ and $q+m_0$ in probability, where m_0 is the smallest integer of m satisfying

$$\omega_m^2 - \omega_{m+1}^2 < 4\ln\zeta_0.$$

Because the second term of $\tilde{\sigma}^2_{k,i}$ has order $O(\ln T/T)$, which is the same order as the penalty term of the BIC, it plays a dominating role. However, because the third term can be at most $O_p\{(\ln T)^{1/2}/T\}$, the approach of $\tilde{p}$ and $\tilde{q}$ to their limits may be slow. The reason that $\tilde{p}$ and $\tilde{q}$ are inconsistent is the long AR order n in the first stage is chosen by the AIC or the BIC. Hannan and Kavalieris (1983b, Theorem 5) have shown that, under some regularity conditions, the MAICE n_A and the MBICE n_B satisfy

$$\begin{aligned}
n_A &= \frac{\ln T}{2\ln\zeta_0}\{1 + o_p(1)\}, \\
n_B &= \frac{\ln T}{2\ln\zeta_0}\{1 + o(1)\}.
\end{aligned}$$

Also, refer to Hannan and Kavalieris (1984b, Theorem 2; 1986a, Theorem 3.3). Thus, the BIC does not penalize overestimation of the orders adequately.

HK-84 have suggested several modifications to correct the bias. One of the easiest ways to obtain consistent estimates of p and q is to strengthen the penalty term of the order selection criterion in the second stage so that

$$P(k,i) = \ln\tilde{\sigma}^2_{k,i} + (k+i)\frac{C(T)}{T},$$

where

$$\lim_{T\to\infty} C(T) = \infty, \quad \lim_{T\to\infty}\frac{\ln T}{C(T)} = 0.$$

In their correction Hannan and Rissanen (1983) have concluded that if the orders $\tilde{p}$ and $\tilde{q}$ are chosen to minimize

$$\mathrm{BIC}^{\delta}(k,i) = \ln \tilde{\sigma}_{k,i}^2 + (k+i)\frac{(\ln T)^{1+\delta}}{T}$$

for fixed δ (> 0), instead of the BIC in the second stage, and if the long AR order n in the stage 1 is $O(\ln T)$, then $\tilde{p} - p$, $\tilde{q} - q$, $\tilde{\phi}_{\tilde{p},j} - \phi_j$, $\tilde{\theta}_{\tilde{q},l} - \theta_l$, $\hat{\phi}_{\tilde{p},j} - \phi_j$, and $\hat{\theta}_{\tilde{q},l} - \theta_l$ converge to 0 a.s. However, this modification is not recommended because it may lead to underestimate the orders for small or moderate T.

If we repeat the second procedure, we can obtain consistent estimates of p and q. In other words, if the second stage is carried out again with $\{\tilde{v}_t\}$ instead of $\{\check{v}_t\}$ and if the resulting estimates of the orders are denoted by $\tilde{p}^{(1)}$ and $\tilde{q}^{(1)}$, then $\tilde{p}^{(1)}$ and $\tilde{q}^{(1)}$ converge respectively to p and q in probability. Denote the corresponding estimates of ϕ_j and θ_l by $\tilde{\phi}_j^{(1)}$ and $\tilde{\theta}_l^{(1)}$, respectively. HK-84 have experienced that if $\tilde{p} = p$ and $\tilde{q} = q$, then $\tilde{p}^{(1)} = p$ and $\tilde{q}^{(1)} = q$ and have concluded that repeating the second stage can hardly fail to improve the estimates. There could be some advantages to repeat the second stage more than once.

There is another adjustment, which results in consistent estimates and is computationally cheaper than one repetition of the second stage. This altered second stage consists of the following three steps.

1. For each $(k,i) \in \{(k,i) \mid k = 0,\ldots,K,\ i = 0,\ldots,I\}$, obtain the OLS estimates $\tilde{\phi}_{k,1},\ldots,\tilde{\phi}_{k,k}$, $\tilde{\theta}_{i,1},\ldots,\tilde{\theta}_{i,i}$ by minimizing

$$\sum_{t=t_0+1}^{T}\left(y_t - \sum_{j=1}^{k}\phi_{k,j}y_{t-j} + \sum_{l=1}^{i}\theta_{i,l}\check{v}_{t-l}\right)^2,$$

where $t_0 = \max(n+k, n+i)$. Then calculate

$$\tilde{\rho}_{k,i}(j) = \sum_{l=0}^{i-j}\tilde{\theta}_{i,l}\tilde{\theta}_{i,l+j},\ j = 0,\ldots,i.$$

2. For each $(k,i) \in \{(k,i) \mid k = 0,\ldots,K,\ i = 0,\ldots,I\}$, calculate $\{\tilde{\omega}_m^2(k,i) \mid m = 0,1,\ldots\}$ by the Levinson-Durbin algorithm.

   (a) As initial values, put

$$a_{k,i}^{(0)}(0) = 1,$$
$$\tilde{\omega}_0^2(k,i) = \tilde{\rho}_{k,i}(0).$$

(b) For $m = 1, 2, \ldots,$

$$a_{k,i}^{(m)}(0) = 1,$$

$$a_{k,i}(m) = \frac{-1}{\tilde{\omega}_{m-1}^2(k,i)} \sum_{l=0}^{m-1} a_{k,i}^{(m-1)}(l)\tilde{\rho}_{k,i}(m-l),$$

$$a_{k,i}^{(m)}(l) = a_{k,i}^{(m-1)}(l) + a_{k,i}(m) a_{k,i}^{(m-1)}(m-l), \quad l = 1, \ldots, m-1,$$

$$\tilde{\omega}_m^2(k,i) = \left\{1 - a_{k,i}^2(m)\right\} \tilde{\omega}_{m-1}^2(k,i).$$

3. For each $(k,i) \in \{(k,i) \mid k = 0, \ldots, K, \ i = 0, \ldots, I\}$, calculate

$$\mathrm{BIC}_{k,i}(r,s) = \begin{cases} \mathrm{BIC}(k,i), & \text{if } r \le k \text{ or } s \le i \\ \mathrm{BIC}(k,i) + n\left\{\tilde{\omega}_0^2(k,i) - \tilde{\omega}_m^2(k,i)\right\}, & \text{if } r > k, \ s > i, \end{cases}$$

where n is the selected order of the long AR model in the first stage and $m = \min(r-k, s-i)$. Then, choose $\tilde{p}_c$ and $\tilde{q}_c$ as the first k and i minimizing $\mathrm{BIC}_{k,i}(r,s)$ at $r = k$ and $s = i$.

Again, $\tilde{p}_c$ and $\tilde{q}_c$ converge to p and q in probability. This approach does not require the condition that $\tilde{\theta}_i(z) = -\sum_{l=0}^{i} \tilde{\theta}_{i,l} z^l$ has all the roots outside the unit circle. Moreover, it is computationally cheap. Hannan and Deistler (1988, p. 254) have chosen $\tilde{p}_c$ and $\tilde{q}_c$ a little bit differently, but the result is the same. They have selected $\tilde{p}_c$ and $\tilde{q}_c$ as the values of k and i, minimizing $\mathrm{BIC}_{k,i}(r,s)$ subject to $k \le H_T$, $i \le H_T$, $r \le H_T$, $s \le H_T$, where $H_T = O\{(\ln T)^a\}$ $(1 < a < \infty)$. We may obtain the minimum of $\mathrm{BIC}_{k,i}(r,s)$ at $(r,s) = (k,i)$. If so, (k,i) is called a "marked" point. Because $(\tilde{p}, \tilde{q})$ is always a marked point and there is at most one other marked point with $k \le \tilde{p}$ and $i \le \tilde{q}$, a lot of calculation is not necessary to find $\tilde{p}_c$ and $\tilde{q}_c$. Clearly, we would rather consider only the case that $k \le \tilde{p}$ and $i \le \tilde{q}$. If we find a marked point, we stop the search for the minimum. Furthermore, it is sufficient to consider $r \le \tilde{p}$ and $s \le \tilde{q}$. It may be applicable to repeat this modified second stage by replacing $\{\tilde{v}_t\}$ with the estimated innovations $\{\tilde{v}_t\}$ calculated from the estimates $\tilde{\phi}_{\tilde{p}_c}(z)$ and $\tilde{\theta}_{\tilde{q}_c}(z)$ of $\phi(z)$ and $\theta(z)$ with $\tilde{p}_c$ and $\tilde{q}_c$. We denote the estimates of p and q obtained through this repeated procedure by $\tilde{p}_c^{(1)}$ and $\tilde{q}_c^{(1)}$. For more details about this modification, refer to Hannan and Deistler (1988, Section 6.5).

Another alternative is to determine the orders in the third stage. Calculate $\{\tilde{v}_t\}$, $\{x_t\}$, and $\{z_t\}$ with $\tilde{p}$ and $\tilde{q}$, i.e.,

$$\tilde{v}_t = y_t = x_t = z_t = 0, \quad t \le 0,$$

$$\tilde{v}_t = y_t - \sum_{j=1}^{\tilde{p}} \tilde{\phi}_{\tilde{p},j} y_{t-j} + \sum_{l=1}^{\tilde{q}} \tilde{\theta}_{\tilde{q},l} \tilde{v}_{t-l}, \quad t = 1, \ldots, T,$$

$$x_t = -\sum_{l=1}^{\tilde{q}} \tilde{\theta}_{\tilde{q},l} x_{t-l} + y_t, \quad t = 1, \ldots, T,$$

$$z_t = -\sum_{l=1}^{\tilde{q}} \tilde{\theta}_{\tilde{q},l} z_{t-l} + \tilde{v}_t, \quad t = 1, \ldots, T.$$

For each $(k,i) \in \{(k,i) \mid k = 0, \ldots, K, \ i = 0, \ldots, I\}$, we obtain an estimate of the innovation variance $\hat{\sigma}^2_{k,i}$ by regressing $\tilde{v}_t + x_t - z_t$ on

$$x_{t-1}, \ldots, x_{t-k}, -z_{t-1}, \ldots, -z_{t-i}$$

for $t = t_1+1, \ldots, T$, where $t_1 = \max(\tilde{p}, \tilde{q})$. Then, choose $\hat{p}$ and $\tilde{q}$ minimizing

$$\text{BIC}(k,i) = \ln \hat{\sigma}^2_{k,i} + (k+i)\frac{\ln T}{T}.$$

The MBICE $\hat{p}$ and $\tilde{q}$ converge to p and q in probability. However, HK-84 have recommended obtaining the orders as proper as possible before going to the third stage.

A modification by Poskitt (1987) also results in consistent estimates of the orders. For each $(k,i) \in \{(k,i) \mid k = 0, \ldots, K, \ i = 0, \ldots, I\}$, obtain the OLS estimates $\tilde{\phi}_{k,1}, \ldots, \tilde{\phi}_{k,k}, \tilde{\theta}_{i,1}, \ldots, \tilde{\theta}_{i,i}$ by minimizing

$$\sum_{t=t_0+1}^{T} \left(y_t - \sum_{j=1}^{k} \phi_{k,j} y_{t-j} + \sum_{l=1}^{i} \theta_{i,l} \check{v}_{t-l} \right)^2,$$

where $t_0 = \max(n+k, n+i)$. Then calculate $\{\tilde{v}_t\}$, $\{x_t\}$, and $\{z_t\}$ by

$$\tilde{v}_t = y_t = x_t = z_t = 0, \quad t \leq 0,$$

$$\tilde{v}_t = y_t - \sum_{j=1}^{k} \tilde{\phi}_{k,j} y_{t-j} + \sum_{l=1}^{i} \tilde{\theta}_{i,l} \tilde{v}_{t-l}, \quad t = 1, \ldots, T,$$

$$x_t = \sum_{j=1}^{k} \tilde{\phi}_{k,j} x_{t-j} + \tilde{v}_t, \quad t = 1, \ldots, T,$$

$$z_t = \sum_{l=1}^{i} \tilde{\theta}_{i,l} z_{t-l} + \tilde{v}_t, \quad t = 1, \ldots, T.$$

Next, obtain the estimates $\tilde{\phi}^{(*)}_{k,1}, \ldots, \tilde{\phi}^{(*)}_{k,k}, \tilde{\theta}^{(*)}_{i,1}, \ldots, \tilde{\theta}^{(*)}_{i,i}$ by minimizing

$$\sum_{t=t_2+1}^{T} \left(\tilde{v}_t + x_t - z_t - \sum_{j=1}^{k} \phi_{k,j} x_{t-j} + \sum_{l=1}^{i} \theta_{i,l} z_{t-l} \right)^2,$$

where $t_2 = \max(k, i)$. Then, calculate the estimated innovations

$$\tilde{v}_t^{(*)} = 0, \quad t \le 0,$$

$$\tilde{v}_t^{(*)} = y_t - \sum_{j=1}^{k} \tilde{\phi}_{k,j}^{(*)} y_{t-j} + \sum_{l=1}^{i} \tilde{\theta}_{i,l}^{(*)} \tilde{v}_{t-l}^{(*)}, \quad t = 1, \ldots, T,$$

and the corresponding estimates of the innovation variance and the Fisher information matrix, which are denoted by $\tilde{\sigma}_{k,i}^{2(*)}$ and $\tilde{\Gamma}_{k,i}(\tilde{v}_t^{(*)})$, respectively. Then, choose $\tilde{p}^*$ and $\tilde{q}^*$, minimizing

$$\Delta(k, i) = T \ln \tilde{\sigma}_{k,i}^{2(*)} + (k+i) \left\{ 1 + \ln \left[\frac{T}{k+i} \operatorname{trace}\left\{ \tilde{\Gamma}_{k,i}^{-1}(\tilde{v}_t^{(*)}) \right\} \right] \right\}.$$

We can heuristically explain the criterion $\Delta(k, i)$ as follows. As k and i increases, $\tilde{\sigma}_{k,i}^{2(*)}$ decreases but it does not change as much when $k > p$ and $i > q$. However, the Fisher information matrix is singular if $k > p$ and $i > q$. Thus, there will be a big difference between $\Delta(k, i)$ and $\Delta(r, s)$, if $r \le p$ or $s \le q$ and if $k > p$ and $i > q$. Poskitt has shown that $\tilde{p}^*$ and $\tilde{q}^*$ are weakly consistent. Of course we can apply this adjusted procedure under the assumption that $k = i$.

Let

$$\boldsymbol{\tau} = (\phi_1, \ldots, \phi_p, \theta_1, \ldots, \theta_q)^t,$$
$$\tilde{\boldsymbol{\tau}} = (\tilde{\phi}_1, \ldots, \tilde{\phi}_p, \tilde{\theta}_1, \ldots, \tilde{\theta}_q)^t,$$
$$\tilde{\boldsymbol{\tau}}^{(1)} = (\tilde{\phi}_1^{(1)}, \ldots, \tilde{\phi}_p^{(1)}, \tilde{\theta}_1^{(1)}, \ldots, \tilde{\theta}_q^{(1)})^t.$$

Then, $\sqrt{T}(\tilde{\boldsymbol{\tau}} - \boldsymbol{\tau})$ has a limiting normal distribution. So does $\sqrt{T}(\tilde{\boldsymbol{\tau}}^{(1)} - \boldsymbol{\tau})$. The covariance matrices of the limiting distributions are complicated, particularly for $\tilde{\boldsymbol{\tau}}^{(1)}$.

As recommended by Newbold and Bos (1983, p. 55), the HR method is very productive and is reasonably easy to apply to practical time series. However, the supporting theory is very profound and sophisticated. Interested readers may consult the references in Section 4.5.

4.3 Koreisha and Pukkila's Method

Recently Koreisha and Pukkila (KP) proposed several linear estimation methods for ARMA modeling (1987, 1988, 1990a, 1990b), which are two-stage methods. The first stage is to obtain the innovation estimates $\{\hat{v}_t\}$ by a long AR model method. The second stage is to apply least squares methods with $\{y_t\}$ and $\{\hat{v}_t\}$ as the regressors. They are fairly similar to the ILS method and the HR method. Because they can be calculated by any regression computer program, no additional programming adjustment is necessary to implement them on any computer system having any basic statistical package.

We first calculate the estimated innovations $\{\hat{v}_t\}$ by fitting a long AR model. More precisely, for large n, we fit the AR(n) model,

$$y_t = \sum_{l=1}^{n} \beta_l y_{t-l} + v_t,$$

to the observations $y_1, \ldots, y_T$. Let $\hat{\beta}_1, \ldots, \hat{\beta}_n$ be the estimates of $\beta_1, \ldots, \beta_n$. Then calculate the estimated innovations

$$\hat{v}_t = -\sum_{l=0}^{n} \hat{\beta}_l y_{t-l}, \quad t = n+1, \ldots, T,$$

where $\hat{\beta}_0 = -1$. KP-90b recommended letting $n = \sqrt{T}$. They used Burg's algorithm to estimate the AR model because it produces less biased estimates than the YW procedure and the estimates are always on the stationarity region. In the second step, we substitute $\hat{v}_t$ for v_t in the ARMA(k, i) model, which yields the regression model

$$y_t - \hat{v}_t = \sum_{j=1}^{k} \phi_{k,j} y_{t-j} - \sum_{l=1}^{i} \theta_{i,l} \hat{v}_{t-l} + u_t,$$

where $\{u_t\}$ is assumed to be a sequence of error terms, and then estimate the parameters preliminarily. However, the estimates may be biased and insufficient because the estimated innovations $\{\hat{v}_t\}$ contain random variations due to estimation errors. Thus, we assume that the true residual term is equal to the estimated innovation plus a white noise term, i.e.,

$$v_t = \hat{v}_t + u_t.$$

Then the regression model becomes

$$y_t - \hat{v}_t = \sum_{j=1}^{k} \phi_{k,j} y_{t-j} - \sum_{l=1}^{i} \theta_{i,l} \hat{v}_{t-l} + u_t - \sum_{l=1}^{i} \theta_{i,l} u_{t-l}.$$

This is a regression equation with MA(i) residual terms. Because the preliminary estimates of the MA parameters are consistent, it is possible to construct an estimated covariance matrix of the error terms; the regression model can then be estimated using a generalized least squares (GLS) method. The GLS estimates are used to update the covariance matrix and then to reestimate the model. This procedure may be repeated until a convergence criterion is satisfied. Because the covariance matrix is banded, its inverse can be easily calculated. (See, e.g. Golub and Van Loan [1989, pp.

149-159].) As mentioned by KP-90a, this approach is analogous to a two-stage limited-information maximum likelihood (LIML) estimation method. Saikkonen (1986) also proposed a GLS method in the second stage.

From some simulation results, KP-90a have observed that the GLS estimates are close to the parameters regardless of sample size and that they are seldom on the noninvertible or the nonstationary regions unlike the OLS and the ML estimates. Then they have concluded that with one iteration and with $n = 0.5\sqrt{T}$ or $\sqrt{T}$ the GLS procedure tends to generate accurate results. KP-90b have also noted that the linear estimate method as well as the HR method yield precise and fast estimates when compared with the number of iterations necessary to calculate the exact ML estimates.

In his letter to the present author, E. J. Hannan has commented as follows.

> There is a virtue in the KP method that might be pointed out. The GLS procedure depends only on an autocovariance sequence and then it does not matter if a root of $\theta(z) = 0$, say z_1, is replaced with $1/\bar{z}_1$. This contrasts with the HR method, which uses filtering to form new variables, and for which it is therefore essential to substitute $1/\bar{z}_1$ for z_1 if $\mid z_1 \mid > 1$. On the other hand the KP method uses the estimated innovation sequence from the initial AR model throughout. Then, I believe, there will be little change from an iteration of the GLS technique since to effect that change one would need improved innovation sequence estimates. As $T \to \infty$, the GLS technique will give efficient estimates. However, for a fixed T, approach to an optimum point may be impossible.

Pukkila, Koreisha, and Kallinen (1990) proposed an ARMA model identification method based on the above linear estimation procedure and a penalty function identification method. Their fundamental idea is similar to the Durbin-Watson statistic (1950, 1951, 1971) and the portmanteau statistic, which will be discussed in Chapter 6. If the orders of the ARMA process are properly selected, then the estimated innovations should be near white noise. To check the randomness of the estimated innovations, they have tested the hypotheses,

$$H_0 : \text{the residuals follow an AR(0) model}$$

versus

$$H_1 : \text{the residuals follow an AR}(r) \text{ model}, \ r > 0,$$

using the BIC and the HQC as test statistics. They have termed this new method the residual white noise autoregressive criterion (RWNAR). If the null hypothesis is rejected, then the chosen orders of the ARMA process being fitted to the data are changed and the residuals associated with the new fitted structure are once again tested for randomness. This iterative process could be terminated, when the residuals of the fitted structure are ascertained to be white noise. For more details about the KP methods, consult the references in Section 4.5.

4.4 The KL Spectral Density

Estimation of parameters and innovations is necessary to determine the orders of an ARMA process by either the HR method or the KP method. In these methods, if the selected orders are judged to be incorrect, we throw away all the information obtained through the procedures and return to the beginning of the identification problem. This is clearly a waste of time and energy. So far as ARMA modeling is concerned, we would rather modify the chosen model to obtain an improved one using the accumulated information, particularly the estimated innovations. This is called the *model modification.* (Refer to Box and Jenkins [1976, pp. 298-299].)

Choi (1990a) has presented the KL spectral density that minimizes the frequency-domain Kullback-Leibler information number. Based on the KL spectral density that is a modification of the initial spectral density, he has suggested an ARMA model modification method. Some numerical examples have illustrated that the KL spectral density estimate and the modified ARMA estimate are superior to the initial estimates, i.e., the autocovariances obtained by the inverse Fourier transform of the KL spectral density estimate are closer to the sample autocovariances of the given observations than those of the initial spectral density estimate.

We denote the spectral density of $\{y_t\}$ and its spectral distribution by

$$S(\lambda) = \frac{1}{2\pi} \sum_{k=-\infty}^{\infty} \sigma(k) \exp(-i\lambda k), \quad -\pi \leq \lambda \leq \pi,$$

$$F(\lambda) = \int_{-\pi}^{\lambda} S(\mu) d\mu, \quad -\pi \leq \lambda \leq \pi,$$

respectively. It is clear (see, e.g., T. W. Anderson [1971, p. 383]) that $F(\lambda)$ is nondecreasing and right-continuous. Also, $F(-\pi) = 0$ and $F(\pi) = \sigma(0)$. Therefore, a spectral distribution has the same properties as a probability distribution up to a constant $\sigma(0)$, and accordingly the spectral density has the same properties as the probability density function up to the constant. In this light some criteria for evaluating closeness of two probability density functions have been applied to measuring the nearness of two spectral densities $S(\lambda)$ and $\check{S}(\lambda)$. For details about the discrimination measures of two spectral densities, readers may consult the references in Section 4.5. Especially we use the frequency-domain Kullback-Leiler information number

$$I(S; \check{S}) = \int_{-\pi}^{\pi} S(\lambda) \ln \frac{S(\lambda)}{\check{S}(\lambda)} d\lambda$$

as a measure to distinguish between two spectral densities $S(\lambda)$ and $\check{S}(\lambda)$. The KL spectral density is defined as the one minimizing $I(S;\check{S})$ subject to the first $p+1$ autocovariance constraints

$$\sigma(0) = \alpha_0, \ldots, \sigma(p) = \alpha_p,$$

where $\alpha_0, \ldots, \alpha_p$ are given constants. Choi (1990a) has shown that the KL spectral density equals

$$S_{KL}(\lambda) = C\frac{\check{S}(\lambda)}{\mid \sum_{k=0}^{p} \phi_k \exp(-i\lambda k) \mid^2}, \quad \phi_0 = -1,$$

where the coefficients $\phi_1, \ldots, \phi_p$ and the positive constant C are the solutions of the simultaneous equations

$$\int_{-\pi}^{\pi} S_{KL}(\lambda) \exp(i\lambda k) d\lambda = \alpha_k, \quad k = 0, \ldots, p.$$

Consider the AR(p) process

$$y_t = \phi_1 y_{t-1} + \cdots + \phi_p y_{t-p} + v_t,$$

where $\{v_t\}$ is a white noise process with variance σ^2. Because the spectral density of the AR(p) process is

$$S_{AR}(\lambda) = \frac{\sigma^2}{2\pi} \frac{1}{\mid \sum_{k=0}^{p} \phi_k \exp(-i\lambda k) \mid^2}, \quad \phi_0 = -1,$$

the KL spectral density is a generalization of a large family of spectral densities including the maximum entropy spectral density and the ARMA spectral density.

Choi (1991c) has presented the following theorem, which gives another rationale for the KL spectral density.

Theorem 4.2. Let $\boldsymbol{v}$ be the n-variate normal random vector with mean $\mathbf{0}$ and covariance matrix $\delta^2 I$. Denote its probability density function by $\phi_n(\boldsymbol{v} \mid \mathbf{0}, \delta^2 I)$. Then, the probability density function minimizing the time-domain Kullback-Leibler information number

$$I(f;\phi_n) = \int f(\boldsymbol{v}) \ln \frac{f(\boldsymbol{v})}{\phi_n(\boldsymbol{v} \mid \mathbf{0}, \delta^2 I)} d\boldsymbol{v}$$

subject to the first $p+1$ autocovariance constraints

$$\sigma(0) = \alpha_0, \ldots, \sigma(p) = \alpha_p$$

is that of $(v_1, \ldots, v_n)^t$, where $\{v_t\}$ is the Gaussian AR(p) process having the same first $p+1$ autocovariances

$$\sigma(0) = \alpha_0, \cdots, \sigma(p) = \alpha_p. \quad \Box$$

This theorem means the Gaussian AR(p) process is the closest in the time-domain Kullback-Leibler sense to independently, identically, and normally distributed random variables subject to the first $p+1$ autocovariance constraints. Thus, if the residuals found from parametric time series modeling or from regression modeling have serially correlated terms, it would be better to regard them as coming from an AR model.

Because the KL spectral density is the closest to the initial spectral density in the time-domain Kullback-Leibler sense as well as in the frequency-domain Kullback-Leibler sense among the ones satisfying the autocovariance constraints, it can be used to modify the initial spectral density estimate and the corresponding time domain model estimate to obtain refined estimates, which are more pertinent to the given observations than the initial ones. Let $\{y_1, \ldots, y_T\}$ be a T-realization of a stationary time series, and let $\{\hat{\sigma}(k)\}$ and $\hat{S}_y(\lambda)$ be estimates of the ACVF and the spectral density, respectively. For a fixed m, if the estimates do not satisfy the relation

$$\int_{-\pi}^{\pi} \hat{S}_y(\lambda) \exp(i\lambda k) d\lambda = \hat{\sigma}(k), \quad k = 0, \ldots, m,$$

even approximately, then we would rather use the KL spectral density estimate

$$\hat{S}_{KL}(\lambda) = \hat{C} \frac{\hat{S}_y(\lambda)}{| \sum_{k=0}^{m} \hat{\phi}_k \exp(-i\lambda k) |^2}, \quad \hat{\phi}_0 = -1$$

instead of the initial estimate $\hat{S}_y(\lambda)$. Here, the coefficients $\hat{\phi}_1, \ldots, \hat{\phi}_m$ and the constant $\hat{C}$ satisfy

$$\int_{-\pi}^{\pi} \hat{S}_{KL}(\lambda) \exp(i\lambda k) d\lambda = \hat{\sigma}(k), \quad k = 0, \ldots, m.$$

Suppose that $\{y_t\}$ can be represented by

$$y_t = f(B) v_t.$$

If $f(z)$ is a rational function of orders $m - p$ (≥ 0) and q (≥ 0) and if $\{v_t\}$ is a white noise process, then $\{y_t\}$ is an ARMA($m - p, q$) process. The KL spectral density of $\{y_t\}$ subject to the first $p + 1$ autocovariance constraints is the spectral density of an ARMA(m, q) process. Its AR coefficients $\pi_1, \ldots, \pi_m$ satisfy the EYW equations

$$\sum_{k=1}^{m} \pi_k \sigma(j-k) = \sigma(j), \quad j = q+1, \ldots, q+m.$$

The corresponding KL parametric model is

$$\pi(B) y_t = f(B) u_t,$$

where $\pi(z) = 1 - \pi_1 z - \cdots - \pi_m z^m$ and $\{u_t\}$ is a white noise process.

If a parametric model is initially used to estimate a spectral density, then there is an easy way to calculate the KL spectral density estimate. Let $\hat{v}_1, \ldots, \hat{v}_T$ be the residuals found from the parametric model associated with the initial spectral density estimate $\hat{S}_y(\lambda)$. If it is a good estimate, then the residuals should not be serially correlated. However, if the sample autocovariances of the residuals are significantly different from 0 for some lags greater than 0, the spectral density estimate should be modified. Assume that p is the largest among the lags whose autocovariances are significantly different from 0. Let

$$\frac{1}{T}\sum_{t=1}^{T} \hat{v}_t^2 = \hat{\alpha}_0, \ldots, \frac{1}{T}\sum_{t=1}^{T-p} \hat{v}_t \hat{v}_{t+p} = \hat{\alpha}_p.$$

Then we calculate the YW estimates $\hat{\phi}_1, \ldots, \hat{\phi}_p$ satisfying

$$\sum_{k=1}^{p} \hat{\phi}_k \hat{\alpha}_{|j-k|} = \hat{\alpha}_j, \quad j = 1, \ldots, p.$$

The KL spectral density estimate is

$$\hat{S}_{KL}(\lambda) = \hat{C}\frac{\hat{S}_y(\lambda)}{|\sum_{k=0}^{p} \hat{\phi}_k \exp(-i\lambda k)|^2}, \quad \hat{\phi}_0 = -1.$$

The related KL estimate of the model is

$$\hat{\phi}(B) y_t = \hat{f}(B) u_t.$$

We can make some conditional probability characterizations of the KL spectral density and the KL model. Interested readers may consult the references in Section 4.5.

Numerical examples illustrate that the KL spectral density estimate and the corresponding KL model estimate are superior to the initial estimates, i.e., the autocovariances of the KL estimates are closer to the sample autocovariances of the given observations than those of the initial estimates. As an example, we consider the uniform random number generator presented by IBM Corporation (1968), which uses a multiplicative-congruential method

$$U_i = 65539 U_{i-1} \mod(2^{31}).$$

If the generator is a good one, the spectral density of the generated numbers will be near a constant over $[-\pi, \pi]$. Otherwise the KL spectral density will result in a better generator. Even though the RANDU has been one of the most widely used subroutines for random number generation, it is well-known (see, e.g., Kennedy and Gentle [1980, pp. 146-148]) that it is not so desirable. In fact, successive triplets tend to satisfy

$$U_i = 6U_{i-1} - 9U_{i-2}.$$

To test the randomness of the RANDU observations, 80 normal random numbers $\{v_t\}$ are generated using the RANDU and Hastings' approximation (1955). Then, it is assumed that the spectral density of $\{v_t\}$ is

$$S_v(\lambda) = \frac{1}{2\pi}, \quad -\pi \le \lambda \le \pi.$$

From the ACRF and the PACF of $\{v_t\}$ we can hardly say that the numbers are random. Moreover, the modified portmanteau statistic is $Q^* = 16.8$ with 12 degrees of freedom. According to Theorem 4.2 the observations may be considered to be from an AR model rather than a white noise model. Several penalty function identification methods such as the FPE, the AIC, the CAT, and the BIC choose 4 as an optimal order. Using the Levinson-Durbin algorithm, we can obtain the KL parametric model estimate and the KL spectral density estimate

$$\hat{\phi}(B)v_t = \hat{u}_t,$$

$$\hat{S}_{KL}(\lambda) = 0.140 \frac{1}{|\hat{\phi}(\exp(-i\lambda))|^2},$$

where

$$\hat{\phi}(z) = 1 + 0.1393z + 0.0441z^2 - 0.0688z^3 + 0.4000z^4.$$

The ACRF and the PACF of the estimated residuals $\{\hat{u}_t\}$ show the randomness. Moreover, the modified portmanteau statistic is $Q^* = 6.93$ with 12 degrees of freedom. Therefore, the KL spectral density estimate, which is the maximum entropy spectral density estimate in this case, and the corresponding AR(4) model estimate are more loyal to the observations $\{v_t\}$ than the initially assumed spectral density and the white noise model. Also, we may conclude that the RANDU is not such an excellent random number generator and the residuals $\{\hat{u}_t\}$ due to the KL parametric model estimate are better random numbers than $\{v_t\}$.

4.5 Additional References

- (Section 4.1) For the inconsistency of the ML estimates in case of $k > p$ and $i > q$, refer to Hannan (1982) and Hannan and Deistler (1988, pp. 186-187).

- (Section 4.1) About some linear estimation methods for a general transfer function model and the ARMAX model, refer to Kashyap and Nasburg (1974), Young (1974, 1984, 1985), Young and Jakeman (1979), Spliid (1983), Stoica, Söderström, Ahlén and Solbrand (1984, 1985), Hannan and Kavalieris (1984a, 1984b, 1986a), Hannan and Deistler (1988), Wahlberg (1989b), and the references therein. Also, consult the references in Section 4.2.

- (Section 4.1) About applications of the long AR model, refer to McClave (1973), Nelson (1974), Graupe, Krause, and Moore (1975), Mentz (1977), Bhansali (1980), An, Chen, and Hannan (1982), and the references therein. As mentioned in Chapter 2, the long AR method has been more popularized since Burg (1967, 1968) proposed the maximum entropy spectral density. Essentially the maximum entropy spectral density is to approximate the nonparametric spectral density by a parametric AR model.

- (Section 4.2) Durbin's idea of innovation regression estimation has been studied by several time series analysts. Kashyap and Nasburg (1974, p. 789) have offered a two-stage least squares procedure. They have commented that the consistency of the estimates should depend critically on the choice of the long AR order n and suggested the use of the FPE. Spliid (1983) provided another dialect of Durbin's method as follows. In the first stage, fit a long AR($k_0 + i_0$) model to the observations, where k_0 and i_0 are fixed. Then, iterate the second stage with the fixed orders k_0 and i_0 until the OLS estimates of the parameters converge. Hannan and McDougall (1988) have compared it with the HR method and have concluded that Spliid's method does not always converge; when it does, it has the same asymptotic distribution as the second stage of the HR method. Wahlberg (1989a) has proposed applying an approximation technique called relative error balanced model reduction (REBMR) to the second stage in order to obtain an efficient ARMA estimate under the assumption that the long AR order n tends to ∞ as T does. Also, refer to Hannan (1980a), Mayne and Firoozan (1982), Åström and Mayne (1983), Saikkonen (1986), and the references therein.

- (Section 4.2) The algebraic structure of Toeplitz matrices has been studied extensively, and many fast algorithms like Levinson's, Durbin's, Wiggins and Robinson's, Trench's, and Zohar's have been presented for solving Toeplitz systems of equations. Roebuck and Barnett (1978) and Demeure and Scharf (1987) presented excellent reviews about the algorithms. A derivation of Algorithm 5.1 can be found in Choi (1990c). Many algorithms similar to Algorithm 5.1 have been proposed by Whittle (1963), Akaike (1973b), Rissanen (1973), Penm and Terrell (1982) and others. However, Algorithm 5.1 is computationally a little bit cheaper than any other one.

- (Section 4.2) The best reference for the HR method is Hannan and Deistler (1988, Chapter 6). Hannan and Kavalieris (1984b) have established the theoretical basis of the HR method more profoundly. There is another modification of the second stage due to Bhansali (1991). For details of the asymptotic covariance matrices of $\sqrt{T}(\tilde{\boldsymbol{\tau}}-\boldsymbol{\tau})$ and $\sqrt{T}\left(\tilde{\boldsymbol{\tau}}^{(1)}-\boldsymbol{\tau}\right)$, refer to Hannan, Dunsmuir, and Deistler (1980) and Chen (1985). Hannan, Kavalieris, and Mackisack (1986) applied the HR method to the ARMAX model. Poskitt (1989) utilized it for transfer function models.

- (Section 4.3) There are more references about the KP method such as Pukkila (1982) and Pukkila and Krishnaiah (1988).
- (Section 4.4) For details about discrepancy measures of two spectral densities, readers may refer to Jenkins and Watts (1968), Robinson (1983), Porat and Friedlander (1984), and the references therein.
- (Section 4.4) Some conditional probability characterizations of the KL spectral density and the KL model can be found in Choi and Cover (1983) and Choi (1985, 1991f).

5

Pattern Identification Methods

Since the early 1980s some methods utilizing the EYW equations have been used for determining the orders of an ARMA process; these are often called the *pattern identification methods.* The penalty function identification methods discussed in Chapter 3 have the advantage of allowing automatic determination of the orders of an ARMA process. However, they are computationally expensive, for they need ML estimates for all possible ARMA models. Even though some innovation regression methods such as the HR and the KP methods in Chapter 4 may be used, they are usually computationally exorbitant. In contrast, the pattern identification methods are computationally cheap.

For a formal introduction to pattern identification methods, we present the EYW equations more precisely.

Property 5.1. A stationary stochastic process has the ARMA(p, q) representation (1.1) *if and only if* the ACVF satisfies a linear difference equation of minimal order p from the minimal rank $q + 1$, i.e., there exist constants $\beta_1, \ldots, \beta_p$ satisfying

$$\sigma(j) = \beta_1 \sigma(j-1) + \cdots + \beta_p \sigma(j-p), \quad j = q+1, q+2, \ldots,$$

where $p > 0$, $\beta_p \neq 0$, $q \geq 0$ and q is the smallest integer satisfying the difference equations. □

In our model, it is clear that $\phi_1 = \beta_1, \ldots, \phi_p = \beta_p$. Property 5.1 implies a rank condition as follows.

Property 5.2. If a stationary stochastic process has the ARMA(p, q) representation (1.1), then, for $k = p, p+1, \ldots,$ $i = q, q+1, \ldots,$ the rank of $B(k, i)$ is $k - \min(k-p, i-q)$. □

The idea of utilizing the EYW equations to determine the orders of an ARMA process is from Bartlett and Diananda (1950). A lot of pattern identification methods have been proposed such as the R and S array method by Gray, Kelley, and McIntire (1978), the Corner method by Beguin, Gourieroux, and Monfort (1980), the GPAC method by Woodward and Gray (1981), another GPAC method by Glasbey (1982), a method using a generalization of both the ACRF and the PACF by Takemura (1984), the ESACF method by Tsay and Tiao (1984), the SCAN method by Tsay and Tiao (1985), the 3-pattern method by Choi (1991a), and so on. There are some references about Properties 5.1 and 5.2 in Section 5.9.

5.1 The 3-Pattern Method

Among various pattern identification methods, we first discuss the 3-pattern method, which uses the patterns of the θ, λ, and η functions. The reason we introduce the three functions prior to the others is the latters can be represented by the formers.

5.1.1 The Three Functions

To introduce the three functions, it is necessary to define some vectors and matrices. Many of them were defined in Section 1.3. For the sake of convenience, we state them again. We first define a k-dimensional Toeplitz matrix and two vectors as

$$B(k,i) = \begin{pmatrix} \rho_i & \rho_{i-1} & \cdots & \rho_{i-k+1} \\ \rho_{i+1} & \rho_i & \cdots & \rho_{i-k+2} \\ \vdots & \vdots & & \vdots \\ \rho_{i+k-1} & \rho_{i+k-2} & \cdots & \rho_i \end{pmatrix},$$

$$\boldsymbol{\rho}(k,i) = (\rho_{i+1}, \ldots, \rho_{i+k})^t,$$
$$\boldsymbol{r}(k,i) = (\rho_{i-k}, \ldots, \rho_{i-1})^t.$$

Also, we denote $\tilde{\boldsymbol{x}} = (x_n, \ldots, x_1)^t$ for any vector $\boldsymbol{x} = (x_1, \ldots, x_n)^t$. For $k = 1, 2, \ldots$ and $i = 0, 1, \ldots,$ if $B(k,i)$ is nonsingular, then we define

$$\begin{aligned} \boldsymbol{\phi}(k,i) &= B^{-1}(k,i)\boldsymbol{\rho}(k,i), \\ \tilde{\boldsymbol{\pi}}(k,i) &= B^{-1}(k,i)\boldsymbol{r}(k,i), \\ \theta(k,i) &= \rho_{i+k+1} - \tilde{\boldsymbol{\phi}}(k,i)^t \boldsymbol{\rho}(k,i), \\ \eta(k,i) &= \rho_{i-k-1} - \boldsymbol{\pi}(k,i)^t \boldsymbol{r}(k,i), \\ \lambda(k,i) &= \rho_i - \boldsymbol{\pi}(k,i)^t \boldsymbol{\rho}(k,i). \end{aligned}$$

For $k = 0$ and $i = 0, 1, \ldots,$ we let

$$\begin{aligned} \theta(0,i) &= \rho_{i+1}, \\ \eta(0,i) &= \rho_{i-1}, \\ \lambda(0,i) &= \rho_i. \end{aligned}$$

We call $\{\theta(k,i)\}$, $\{\lambda(k,i)\}$, and $\{\eta(k,i)\}$ the θ, λ, and η functions, respectively. We denote the jth elements of $\boldsymbol{\phi}(k,i)$ and $\boldsymbol{\pi}(k,i)$ by $\phi_{k,j}^{(i)}$ and $\pi_{k,j}^{(i)}$

for $j = 1, \ldots, k$. Also, let $\phi_{k,0}^{(i)} = \pi_{k,0}^{(i)} = -1$ for each pair (k, i). Because these definitions are the same as the ones in Algorithm 1.2, we can calculate the θ, λ, and η functions recursively using the simplified Trench-Zohar algorithm. We also define a function and two index sets as

$$\begin{aligned}
\phi_k^{(i)}(z) &= -\phi_{k,0}^{(i)} - \phi_{k,1}^{(i)} z - \cdots - \phi_{k,k}^{(i)} z^k, \quad \phi_{k,0}^{(i)} = -1, \\
I_{r,s} &= \{(k, s) \mid k = r, r+1, \ldots\} \bigcup \{(r, i) \mid i = s, s+1, \ldots\}, \\
J_{r,s} &= \{(k, i) \mid k = r, r+1, \ldots, \ i = s, s+1, \ldots\}.
\end{aligned}$$

If $k > p$ and $i > q$, then $B(k, i)$ is singular, and then the three functions are not defined. However, if the inverse of $B(k, i)$ is taken as a generalized inverse, then the three functions can be extended as follows.

$$\begin{aligned}
\theta^-(k, i) &= \rho_{i+k+1} - \tilde{\boldsymbol{\rho}}(k, i)^t B^-(k, i) \boldsymbol{\rho}(k, i), \\
\eta^-(k, i) &= \rho_{i-k-1} - \tilde{\boldsymbol{r}}(k, i)^t B^-(k, i) \boldsymbol{r}(k, i), \\
\lambda^-(k, i) &= \rho_i - \tilde{\boldsymbol{\rho}}(k, i)^t B^-(k, i) \boldsymbol{r}(k, i).
\end{aligned}$$

Choi (1991a) has derived the patterns of the three functions using the EYW equations, which are useful for ARMA model identification. However, we shall derive them more intuitively.

We know that

$$\sigma(0)\theta(k, i) = \text{cov}(y_t, \phi_k^{(i)}(B) y_{t+k+i+1}).$$

If $(k, i) \in I_{p,q}$, then it equals

$$\text{cov}(y_t, \phi(B) y_{t+k+i+1}) = \text{cov}(y_t, \theta(B) v_{t+k+i+1}),$$

which is 0 by the stationarity assumption.

Property 5.3. *The θ Pattern*
A stochastic process has the ARMA(p, q) representation (1.1) *if and only if* p and q are the smallest integers satisfying

$$\theta(k, i) = 0, \quad (k, i) \in I_{p,q}. \quad \square$$

Property 5.4. *The Generalized θ Pattern*
If a stochastic process has the ARMA(p, q) representation (1.1), then

$$\theta^-(k, i) = 0, \quad (k, i) \in J_{p,q}. \quad \square$$

The two-way array of the θ^- function is tabulated in Table 5.1.

We can easily show that

$$\lambda(k, i) = \rho_i - \tilde{\boldsymbol{\phi}}(k, i)^t \boldsymbol{r}(k, i),$$

Table 5.1. The θ^- array

$k \setminus i$	$\cdots$	$q-1$	q	$q+1$	$\cdots$
$\vdots$		$\vdots$	$\vdots$	$\vdots$	
$p-1$	$\cdots$	$\theta(p-1,q-1)$	$\theta(p-1,q)$	$\theta(p-1,q+1)$	$\cdots$
p	$\cdots$	$\theta(p,q-1)$	0	0	$\cdots$
$p+1$	$\cdots$	$\theta(p+1,q-1)$	0	0	$\cdots$
$p+2$	$\cdots$	$\theta(p+2,q-1)$	0	0	$\cdots$
$\vdots$		$\vdots$	$\vdots$	$\vdots$	

which implies that

$$\sigma(0)\lambda(k,i) = \text{cov}(y_t, \phi_k^{(i)}(B)y_{t+i}).$$

If $k = p$ and $i > q$, then

$$\begin{aligned} \sigma(0)\lambda(p,i) &= \text{cov}(y_t, \phi(B)y_{t+i}) \\ &= \text{cov}(y_t, \theta(B)v_{t+i}), \end{aligned}$$

which is 0. If $k > p$ and $i = q$, then

$$\begin{aligned} \sigma(0)\lambda(k,q) &= \text{cov}(y_t, \phi(B)y_{t+q}) \\ &= \text{cov}(y_t, \theta(B)v_{t+q}) \\ &= \text{cov}(y_t, -\theta_q v_t) \\ &= \theta_0\theta_q\sigma^2. \end{aligned}$$

Property 5.5. *The λ Pattern*
A stochastic process has the ARMA(p,q) representation (1.1) *if and only if* p and q are the smallest integers satisfying

$$\begin{aligned} &\lambda(p,q+1) = \lambda(p,q+2) = \cdots = 0, \\ &\lambda(p,q) = \lambda(p+1,q) = \cdots = \frac{\theta_0\theta_q\sigma^2}{\sigma(0)}. \quad \square \end{aligned}$$

If the underlying process is from a pure AR model, i.e., $q = 0$, then

$$\lambda(p) = \lambda(p+1) = \cdots = \frac{\sigma^2}{\sigma(0)}.$$

We can estimate the innovation variance σ^2 as

$$\hat{\sigma}_k^2 = \hat{\sigma}(0)\hat{\lambda}(k), \quad k = p, p+1, \ldots.$$

Table 5.2. The λ^- array

$k \setminus i$	$\cdots$	q	$q+1$	$q+2$	$\cdots$
$\vdots$		$\vdots$	$\vdots$	$\vdots$	
$p-1$	$\cdots$	$\lambda(p-1,q)$	$\lambda(p-1,q+1)$	$\lambda(p-1,q+2)$	$\cdots$
p	$\cdots$	$\lambda(p,q)$	0	0	$\cdots$
$p+1$	$\cdots$	$\lambda(p,q)$	X	0	$\cdots$
$p+2$	$\cdots$	$\lambda(p,q)$	X	X	$\cdots$
$\vdots$		$\vdots$	$\vdots$	$\vdots$	

Note: X represents nonzero.

The estimates may be used to determine the AR order by the penalty function methods. Penm and Terrell (1982) applied this idea to vector AR model identification.

Property 5.6. *The Generalized λ Pattern*
If a stochastic process has the ARMA(p,q) representation (1.1), then the following holds.

1. If $i > q$, then $\lambda^-(p,i) = 0$.
2. If $k \geq p$, then $\lambda^-(k,q) = \theta_0\theta_q\sigma^2/\sigma(0)$.
3. If $s > r \geq 0$, then $\lambda^-(p+r, q+s) = 0$. □

The two-way array of the λ^- function is tabulated in Table 5.2.

Consider an ARMA(p,q) model, which is associated with the ARMA(p, q) model (1.1),

$$\pi(B^{-1})z_t = \theta(B^{-1})v_t,$$

where

$$\pi(z) = \frac{1}{\phi_p} z^p \phi(z^{-1}) = -\pi_0 - \pi_1 z - \cdots - \pi_p z^p, \quad \pi_0 = -1.$$

If the spectral densities of $\{y_t\}$ and $\{z_t\}$ are denoted by $S_y(\lambda)$ and $S_z(\lambda)$, respectively, then

$$S_z(\lambda) = \phi_p^2 S_y(\lambda).$$

The Wiener-Khinchine relation implies the two processes have the common ACRF $\{\rho_j\}$. Because all the roots of $\phi(z) = 0$ are outside the unit circle, all the roots of $\pi(z) = 0$ are inside the unit circle. Thus, the ACRF satisfies the backward EYW equations

$$\rho_j = \pi_1\rho_{j+1} + \cdots + \pi_p\rho_{j+p}, \quad j = q-p+1, q-p+2, \ldots,$$

which yield

$$\tilde{\boldsymbol{\pi}}(p,q) = B^{-1}(p,q)\boldsymbol{r}(p,q).$$

Table 5.3. The η^- array

$k \setminus i$	$\cdots$	$q+1$	$q+2$	$q+3$	$\cdots$
$\vdots$		$\vdots$	$\vdots$	$\vdots$	
$p-1$	$\cdots$	$\eta(p-1,q+1)$	$\eta(p-1,q+2)$	$\eta(p-1,q+3)$	$\cdots$
p	$\cdots$	$\eta(p,q+1)$	0	0	$\cdots$
$p+1$	$\cdots$	∞	$\eta(p,q+1)$	0	$\cdots$
$p+2$	$\cdots$	∞	∞	$\eta(p,q+1)$	$\cdots$
$\vdots$		$\vdots$	$\vdots$	$\vdots$	

Therefore, $\pi_{p,j}^{(q)} = \pi_j$ for $j = 1, \ldots, p$. The backward EYW equations imply the following.

Property 5.7. *The η Pattern*
If a stochastic process has the ARMA(p, q) representation (1.1), then the following holds.

1. $\eta(p-1, i) \neq 0, \quad i = q+1, q+2, \ldots.$
2. $\eta(p, q+1) = -\lambda(p, q)/\phi_p.$
3. $\eta(p, q+2) = \eta(p, q+3) = \cdots = 0. \quad \square$

Property 5.8. *The Generalized η Pattern*
If a stochastic process has the ARMA(p, q) representation (1.1), then the following holds.

1. If $r \geq 0$, then $\eta^-(p+r, q+1+r) = -\lambda(p, q)/\phi_p.$
2. If $r > s \geq 0$, then $\eta^-(p+r, q+1+s) = \pm\infty.$
3. If $s > r \geq 0$, then $\eta^-(p+r, q+1+s) = 0. \quad \square$

The two-way array of the η^- function is tabulated in Table 5.3.

For fixed T, $\hat{B}(k, i)$ is nonsingular with probability 1. Thus, $\hat{\theta}^-(k, i)$, $\hat{\lambda}^-(k, i)$, and $\hat{\eta}^-(k, i)$ equal $\hat{\theta}(k, i)$, $\hat{\lambda}(k, i)$, and $\hat{\eta}(k, i)$, respectively with probability 1. Thus, in practice we can apply Properties 5.4, 5.6, and 5.8 to $\hat{\theta}(k, i)$, $\hat{\lambda}(k, i)$ and $\hat{\eta}(k, i)$, respectively.

5.1.2 Asymptotic Distributions

Visual inspection of the $\hat{\theta}$, $\hat{\lambda}$, and $\hat{\eta}$ arrays may give an answer to selecting the orders of the ARMA process. However, to determine the orders statistically using the three functions, we need their distributions. Choi (1990g, 1991a) presented their asymptotic distributions.

Theorem 5.1. *The Asymptotic Distribution of the θ Array*
Let $\{y_1, \ldots, y_T\}$ be a T-realization from the ARMA(p,q) model (1.1). If $(k,i) \in I_{p,q}$, then $\sqrt{T}\hat{\theta}(k,i)$'s are asymptotically normally distributed with means 0 and covariances

$$\lim_{T\to\infty} \operatorname{cov}\Big(\sqrt{T}\hat{\theta}(k,i), \sqrt{T}\hat{\theta}(j,l)\Big)$$
$$= \frac{\sigma^2}{\sigma^2(0)} \sum_{r=0}^{p}\sum_{s=0}^{q}\sum_{m=0}^{p}\sum_{n=0}^{q} \phi_r\theta_s\phi_m\theta_n\sigma(k+i-j -l-r-s+m+n). \quad \square$$

Theorem 5.2. *The Asymptotic Distribution of the λ Array*
Let $\{y_1, \ldots, y_T\}$ be a T-realization from the ARMA(p,q) model (1.1). Then $\sqrt{T}\hat{\sigma}(0)\hat{\lambda}(p,q+1), \sqrt{T}\hat{\sigma}(0)\hat{\lambda}(p,q+2), \ldots$ and $\sqrt{T}\left\{\hat{\sigma}(0)\hat{\lambda}(p,q) - \sigma(0)\lambda(p,q)\right\}$, $\sqrt{T}\left\{\hat{\sigma}(0)\hat{\lambda}(p+1,q) - \sigma(0)\lambda(p+1,q)\right\}, \ldots$ are asymptotically normally distributed with means 0. The asymptotic covariances are as follows.

If $i,j = q+1, q+2, \ldots$, then

$$\lim_{T\to\infty} \operatorname{cov}\left(\sqrt{T}\hat{\sigma}(0)\hat{\lambda}(p,i), \sqrt{T}\hat{\sigma}(0)\hat{\lambda}(p,j)\right)$$
$$= \frac{\sigma^2}{\phi_p^2} \sum_{r=0}^{p}\sum_{s=0}^{q}\sum_{m=0}^{p}\sum_{n=0}^{q} \phi_r\theta_s\phi_m\theta_n\sigma(i-j-r-s+m+n).$$

If $k = p, p+1, \ldots, i = q+1, q+2, \ldots$, then

$$\lim_{T\to\infty} \operatorname{cov}\left(\sqrt{T}\left\{\hat{\sigma}(0)\hat{\lambda}(k,q) - \sigma(0)\lambda(k,q)\right\}, \sqrt{T}\hat{\sigma}(0)\hat{\lambda}(p,i)\right)$$
$$= -\frac{\sigma^2}{\phi_p} \sum_{r=0}^{k}\sum_{s=0}^{q}\sum_{m=0}^{p}\sum_{n=0}^{q} \pi_{k,r}^{(q)}\theta_s\phi_m\theta_n\sigma(q-p-i+r-s+m+n).$$

If $E(v_t^4) < \infty$, then, for $k, j = p, p+1, \ldots,$

$$\lim_{T\to\infty} \operatorname{cov}\left(\sqrt{T}\left\{\hat{\sigma}(0)\hat{\lambda}(k,q) - \sigma(0)\lambda(k,q)\right\}, \sqrt{T}\left\{\hat{\sigma}(0)\hat{\lambda}(j,q) - \sigma(0)\lambda(j,q)\right\}\right)$$
$$= \sigma^2 \sum_{r=0}^{k}\sum_{s=0}^{q}\sum_{m=0}^{j}\sum_{n=0}^{q} \pi_{k,r}^{(q)}\theta_s\pi_{j,m}^{(q)}\theta_n\sigma(r-s-m+n) + \theta_q^2\{E(v_t^4) - 2\sigma^4\}. \quad \square$$

In the following theorem we will relax the fourth-moment condition, i.e., $E(v_t^4) < \infty$. The coefficient ψ_j in the theorem is the jth MA coefficient of the MA(∞) representation of the ARMA(p, q) model as defined in Section 1.3.

Theorem 5.3. *The Asymptotic Distribution of the λ Array*
Let $\{y_1, \ldots, y_T\}$ be a T-realization from the ARMA(p, q) model (1.1). Then, $\sqrt{T}\hat{\lambda}(p, q+1)$, $\sqrt{T}\hat{\lambda}(p, q+2)$, $\ldots$ and $\sqrt{T}\left\{\hat{\lambda}(p,q) - \lambda(p,q)\right\}$, $\sqrt{T}\left\{\hat{\lambda}(p+1,q) - \lambda(p+1,q)\right\}, \ldots$ are asymptotically normally distributed with means 0. The asymptotic covariances are as follows.

If $i, j = q+1, q+2, \ldots$, then

$$\lim_{T\to\infty} \operatorname{cov}\left(\sqrt{T}\hat{\lambda}(p,i), \sqrt{T}\hat{\lambda}(p,j)\right)$$
$$= \frac{\sigma^2}{\sigma^2(0)\phi_p^2} \sum_{r=0}^{p}\sum_{s=0}^{q}\sum_{m=0}^{p}\sum_{n=0}^{q} \phi_r\theta_s\phi_m\theta_n\sigma(i-j-r-s+m+n).$$

If $k, j = p, p+1, \ldots$, then

$$\lim_{T\to\infty} \operatorname{cov}\left(\sqrt{T}\left\{\hat{\lambda}(k,q) - \lambda(k,q)\right\}, \sqrt{T}\left\{\hat{\lambda}(j,q) - \lambda(j,q)\right\}\right)$$
$$= \frac{\sigma^2}{\sigma^2(0)} \sum_{r=0}^{k}\sum_{s=0}^{q}\sum_{\alpha=0}^{j}\sum_{\beta=0}^{q} \pi_{k,r}^{(q)}\theta_s\pi_{j,\alpha}^{(q)}\theta_\beta\sigma(r-s-\alpha+\beta)$$
$$-2\lambda(p,q)\frac{\sigma^2}{\sigma^2(0)} \sum_{r=0}^{k}\sum_{s=0}^{q}\sum_{\delta=0}^{\infty} \pi_{k,r}^{(q)}\theta_s\psi_\delta\sigma(q+r-s+\delta)$$
$$-2\lambda(p,q)\frac{\sigma^2}{\sigma^2(0)} \sum_{\alpha=0}^{j}\sum_{\beta=0}^{q}\sum_{m=0}^{\infty} \pi_{j,\alpha}^{(q)}\theta_\beta\psi_m\sigma(q+\alpha-\beta+m)$$
$$+2\frac{\lambda^2(p,q)}{\sigma^2(0)} \sum_{r=-\infty}^{\infty} \sigma^2(r) + \lambda^2(p,q).$$

If $k = p, p+1, \ldots, i = q+1, q+2, \ldots$, then

$$\lim_{T\to\infty} \operatorname{cov}\left(\sqrt{T}\left\{\hat{\lambda}(k,q) - \lambda(k,q)\right\}, \sqrt{T}\hat{\lambda}(p,i)\right)$$
$$= -\frac{\sigma^2}{\phi_p\sigma^2(0)} \sum_{r=0}^{k}\sum_{s=0}^{q}\sum_{\alpha=0}^{p}\sum_{\beta=0}^{q} \pi_{k,r}^{(q)}\theta_s\phi_\alpha\theta_\beta\sigma(q+r-s-i-p+\alpha+\beta)$$
$$+\frac{2\lambda(p,q)\sigma^2}{\phi_p\sigma^2(0)} \sum_{\alpha=0}^{p}\sum_{\beta=0}^{q}\sum_{m=0}^{\infty} \phi_\alpha\theta_\beta\psi_m\sigma(i+p-\alpha-\beta+m). \quad \square$$

Theorem 5.4. *The Asymptotic Distribution of the η Array*
Let $\{y_1, \ldots, y_T\}$ be a T-realization from the ARMA(p, q) model (1.1). Then, $\sqrt{T}\left\{\hat{\sigma}(0)\hat{\phi}_{p,p}^{(q)}\hat{\eta}(p,q+1) - \sigma(0)\phi_p\eta(p,q+1)\right\}$, $\sqrt{T}\hat{\sigma}(0)\hat{\phi}_{p,p}^{(q+1)}\hat{\eta}(p,$

$q+2$), $\sqrt{T}\hat{\sigma}(0)\hat{\phi}_{p,p}^{(q+2)}\hat{\eta}(p,q+3),\ldots$ are asymptotically normally distributed with means 0. The asymptotic covariances are as follows.

If $i, j = q+2, q+3, \ldots$, then

$$\lim_{T\to\infty} \operatorname{cov}\left(\sqrt{T}\hat{\sigma}(0)\hat{\phi}_{p,p}^{(i-1)}\hat{\eta}(p,i), \sqrt{T}\hat{\sigma}(0)\hat{\phi}_{p,p}^{(j-1)}\hat{\eta}(p,j)\right)$$
$$= \frac{\sigma^2}{\phi_p^2}\sum_{r=0}^{p}\sum_{s=0}^{q}\sum_{m=0}^{p}\sum_{n=0}^{q}\phi_r\theta_s\phi_m\theta_n\sigma(i-j-r-s+m+n).$$

If $i = q+2,\ q+3,\ldots$, then

$$\lim_{T\to\infty} \operatorname{cov}\left(\sqrt{T}\left\{\hat{\sigma}(0)\hat{\phi}_{p,p}^{(q)}\hat{\eta}(p,q+1) - \sigma(0)\phi_p\eta(p,q+1)\right\},\right.$$
$$\left.\sqrt{T}\hat{\sigma}(0)\hat{\phi}_{p,p}^{(i-1)}\hat{\eta}(p,i)\right)$$
$$= -\frac{\sigma^2}{\phi_p}\sum_{r=0}^{p}\sum_{s=0}^{q}\sum_{m=0}^{p}\sum_{n=0}^{q}\phi_r\theta_s\pi_m\theta_n\sigma(p+i-1-q-r-s-m+n).$$

If $E(v_t^4) < \infty$, then for $k = p, i = q+1$,

$$\lim_{T\to\infty} \operatorname{var}\left(\sqrt{T}\left\{\hat{\sigma}(0)\hat{\phi}_{p,p}^{(q)}\hat{\eta}(p,q+1) - \sigma(0)\phi_p\eta(p,q+1)\right\}\right)$$
$$= \sigma^2\sum_{r=0}^{p}\sum_{s=0}^{q}\sum_{m=0}^{p}\sum_{n=0}^{q}\pi_r\theta_s\pi_m\theta_n\sigma(r-s-m+n)$$
$$+\theta_q^2\left\{E(v_t^4) - 2\sigma^4\right\}. \quad \square$$

As in Theorem 5.3, we can relax the fourth-order moment condition of Theorem 5.4.

Theorem 5.5. *The Asymptotic Distribution of the η Array*
Let $\{y_1,\ldots,y_T\}$ be a T-realization from the ARMA(p,q) model (1.1). Then, $\sqrt{T}\left\{\hat{\phi}_{p,p}^{(q)}\hat{\eta}(p,q+1) - \phi_p\eta(p,q+1)\right\}$, $\sqrt{T}\hat{\phi}_{p,p}^{(q+1)}\hat{\eta}(p,q+2)$, $\sqrt{T}\hat{\phi}_{p,p}^{(q+2)}$ $\hat{\eta}(p,q+3),\ldots$ are asymptotically normally distributed with means 0. The asymptotic covariances are as follows.

If $i, j = q+2, q+3, \ldots$, then

$$\lim_{T\to\infty} \operatorname{cov}\left(\sqrt{T}\hat{\phi}_{p,p}^{(i-1)}\hat{\eta}(p,i), \sqrt{T}\hat{\phi}_{p,p}^{(j-1)}\hat{\eta}(p,j)\right)$$
$$= \frac{\sigma^2}{\sigma^2(0)\phi_p^2}\sum_{r=0}^{p}\sum_{s=0}^{q}\sum_{m=0}^{p}\sum_{n=0}^{q}\phi_r\theta_s\phi_m\theta_n\sigma(i-j-r-s+m+n).$$

If $i = q+2, q+3, \ldots$, then

$$\lim_{T\to\infty} \operatorname{cov}\left(\sqrt{T}\left\{\hat{\phi}_{p,p}^{(q)}\hat{\eta}(p,q+1) - \phi_p\eta(p,q+1)\right\}, \sqrt{T}\hat{\phi}_{p,p}^{(i-1)}\hat{\eta}(p,i)\right)$$

$$= -\frac{\sigma^2}{\phi_p \sigma^2(0)} \sum_{r=0}^{p} \sum_{s=0}^{q} \sum_{\alpha=0}^{p} \sum_{\beta=0}^{q} \pi_r \theta_s \phi_\alpha \theta_\beta \times \sigma(q + r - s - i + 1 - p + \alpha + \beta) + \frac{2\lambda(p,q)\sigma^2}{\phi_p \sigma^2(0)} \sum_{\alpha=0}^{p} \sum_{\beta=0}^{q} \sum_{m=0}^{\infty} \phi_\alpha \theta_\beta \psi_m \sigma(i - 1 + p - \alpha - \beta + m).$$

If $k = p, i = q + 1$, then

$$\lim_{T\to\infty} \operatorname{var}\left(\sqrt{T}\left\{\hat{\phi}_{p,p}^{(q)} \hat{\eta}(p, q+1) - \phi_p \eta(p, q+1)\right\}\right) = \frac{\sigma^2}{\sigma^2(0)} \sum_{r=0}^{p} \sum_{s=0}^{q} \sum_{\alpha=0}^{p} \sum_{\beta=0}^{q} \pi_r \theta_s \pi_\alpha \theta_\beta \sigma(r - s - \alpha + \beta) - 4\lambda(p,q) \frac{\sigma^2}{\sigma^2(0)} \sum_{r=0}^{p} \sum_{s=0}^{q} \sum_{m=0}^{\infty} \pi_r \theta_s \psi_m \sigma(q + r - s + m) + 2\frac{\lambda^2(p,q)}{\sigma^2(0)} \sum_{r=-\infty}^{\infty} \sigma^2(r) + \lambda^2(p,q). \quad \square$$

Theorem 5.6. *The Asymptotic Distribution of the η Array*
Let $\{y_1, \ldots, y_T\}$ be a T-realization from the ARMA(p, q) model (1.1). Denote the (m, n) element of $B^{-1}(p, q)$ by $\rho^{m,n}$. Then, $\sqrt{T}\{\hat{\eta}(p, q + 1) - \eta(p, q + 1)\}$, $\sqrt{T}\hat{\eta}(p, q + 2)$, $\sqrt{T}\hat{\eta}(p, q + 3), \ldots$ are asymptotically normally distributed with means 0. The asymptotic covariances are as follows.

If $i, j = q + 2, q + 3, \ldots$, then

$$\lim_{T\to\infty} \operatorname{cov}\left(\sqrt{T}\hat{\eta}(p, i), \sqrt{T}\hat{\eta}(p, j)\right) = \frac{\sigma^2}{\sigma^2(0)\phi_p^4} \sum_{r=0}^{p} \sum_{s=0}^{q} \sum_{m=0}^{p} \sum_{n=0}^{q} \phi_r \theta_s \phi_m \theta_n \sigma(i - j - r - s + m + n).$$

If $i = q + 2, q + 3, \ldots$, then

$$\lim_{T\to\infty} \operatorname{cov}\left(\sqrt{T}\left\{\hat{\eta}(p, q+1) - \eta(p, q+1)\right\}, \sqrt{T}\hat{\eta}(p, i)\right) = -\frac{\sigma^2}{\phi_p^4 \sigma^2(0)} \sum_{r=0}^{p} \sum_{s=0}^{q} \sum_{\alpha=0}^{p} \sum_{\beta=0}^{q} (\phi_p \pi_r + \lambda(p,q)\rho^{p,r}) \times \theta_s \phi_\alpha \theta_\beta \sigma(q - i + 1 - p + r - s + \alpha + \beta) + \frac{2\lambda(p,q)\sigma^2}{\phi_p^3 \sigma^2(0)} \sum_{\alpha=0}^{p} \sum_{\beta=0}^{q} \sum_{m=0}^{\infty} \phi_\alpha \theta_\beta \psi_m \sigma(i - 1 + p - \alpha - \beta + m).$$

If we let $\rho^{p,0} = 0$, then for $k = p, i = q + 1$n,

$$\lim_{T\to\infty} \operatorname{var}\left(\sqrt{T}\,\{\hat{\eta}(p,q+1) - \eta(p,q+1)\}\right)$$

$$= \frac{\sigma^2}{\phi_p^4\sigma^2(0)} \sum_{r=0}^{p}\sum_{s=0}^{q}\sum_{\alpha=0}^{p}\sum_{\beta=0}^{q} \theta_s\theta_\beta(\phi_p\pi_r + \lambda(p,q)\rho^{p,r})$$
$$\times(\phi_p\pi_\alpha + \lambda(p,q)\rho^{p,\alpha})\sigma(r-s-\alpha+\beta)$$
$$-\frac{4\lambda(p,q)\sigma^2}{\phi_p^3\sigma^2(0)} \sum_{r=0}^{p}\sum_{s=0}^{q}\sum_{\delta=0}^{\infty}(\phi_p\pi_r + \lambda(p,q)\rho^{p,r})\theta_s\psi_\delta\sigma(q+r-s+\delta)$$
$$+\frac{2\lambda^2(p,q)}{\phi_p^2\sigma^2(0)} \sum_{m=-\infty}^{\infty} \sigma^2(m) + \frac{\lambda^2(p,q)}{\phi_p^2}. \quad \square$$

5.1.3 Two Chi-Squared Statistics

Property 5.1 implies that the following hypotheses are equivalent.

H_K : the true orders are p and q.
$H_E : \theta(p,q-1) \neq 0$, $\theta(p-1,q) \neq 0$ and $\theta(p,i) = 0$ for $i = q, q+1, \ldots$.
$H_J : \theta(p-1,q) \neq 0$, $\theta(p,q-1) \neq 0$ and $\theta(k,q) = 0$ for $k = p, p+1, \ldots$.

Using the asymptotic distributions of Theorem 5.1, we can derive test statistics to test the hypotheses H_E and H_J. Let

$$\boldsymbol{h}_r(k,i) = \left(\hat{\theta}(k,i), \hat{\theta}(k,i+1), \ldots, \hat{\theta}(k,i+r-1)\right)^t,$$
$$\boldsymbol{v}_r(k,i) = \left(\hat{\theta}(k,i), \hat{\theta}(k+1,i), \ldots, \hat{\theta}(k+r-1,i)\right)^t,$$

and let $W_r(k,i)$ be an r-dimensional symmetric Toeplitz matrix whose (α,β) element is

$$\frac{1}{\hat{\sigma}^2(0)} \sum_{m=-i}^{i} \hat{\sigma}_z(m)\hat{\sigma}_z(m+\alpha-\beta),$$

where

$$\hat{\sigma}_z(m) = \sum_{r=0}^{k}\sum_{s=0}^{k} \hat{\phi}_{k,r}^{(i)}\hat{\phi}_{k,s}^{(i)}\hat{\sigma}(m+r-s).$$

Choi (1990h) have defined the E and J statistics[1] as

$$E_r(k,i) = T\boldsymbol{h}_r(k,i)^t W_r^{-1}(k,i)\boldsymbol{h}_r(k,i),$$
$$J_r(k,i) = T\boldsymbol{v}_r(k,i)^t W_r^{-1}(k,i)\boldsymbol{v}_r(k,i).$$

[1]The E and J statistics were named after the late *EunJung Kim*.

Theorem 5.1 yields their asymptotic distributions as follows.

Corollary 5.1. Let $\{y_1, \ldots, y_T\}$ be a T-realization from the ARMA(p, q) model (1.1). If $k = p$ and $i = q, q+1, \ldots,$ then $E_r(k, i)$ is asymptotically χ_r^2 distributed. If $k = p, p+1, \ldots$ and $i = q$, then $J_r(k, i)$ is asymptotically χ_r^2 distributed. □

Theoretically speaking, we should use $E_\infty(k, i)$ and $J_\infty(k, i)$ to test the null hypotheses H_E and H_J. However, there are only a finite number of observations. Moreover, because the behavior of $\hat{\rho}_j$ becomes more erratic as $\mid j \mid$ increases, the asymptotic distributions of E_r and J_r in Corollary 5.1 may be different from their empirical ones if r is too large. Thus, we will test the following modified hypotheses

$$H_E(r) : \theta(p, q-1) \neq 0, \theta(p-1, q) \neq 0 \quad \text{and} \quad \begin{array}{l} \theta(p, i) = 0 \\ \text{for } i = q, \ldots, q+r-1, \end{array}$$

$$H_J(r) : \theta(p-1, q) \neq 0, \theta(p, q-1) \neq 0 \quad \text{and} \quad \begin{array}{l} \theta(k, q) = 0 \\ \text{for } k = p, \cdots, p+r-1, \end{array}$$

with moderate r. The present author made a subjective conclusion based on simulation results that it is good enough for ARMA modeling to test $H_E(3)$ and $H_J(3)$.

We summarize the strategy of determining the orders of the ARMA process using the 3-patterns and the two χ^2 statistics as follows.

1. Estimate the θ, λ, and η arrays. Choose possible pairs (k, i) of the orders based on Properties 5.3-5.8.

2. Choose the smallest k and i among the possible pairs satisfying

$$E_r(k, i) < \chi_r^2(1-\alpha),$$
$$J_r(k, i) < \chi_r^2(1-\alpha),$$

   where $\chi_r^2(1-\alpha)$ is the $(1-\alpha)$th quantile point of the χ_r^2 random variable.

3. Confirm the selection by checking that

$$E_r(k, n) < \chi_r^2(1-\alpha), \; n = i+1, i+2, \ldots,$$
$$J_r(m, i) < \chi_r^2(1-\alpha), \; m = k+1, k+2, \ldots,$$

   and that $E_r(k, i-1)$ and $J_r(k-1, i)$ are significantly larger than 0.

In this section we used only the asymptotic distribution of the θ function. We can make more χ^2-type test statistics based on the asymptotic distributions of the other two functions.

5.2 The R and S Array Method

5.2.1 The R and S Patterns

Gray, Kelley, and McIntire (1978) (GKM) have presented the *R and S array method.*

Definition 5.1. For an integer m, a positive constant h and a real-valued function f, we let $f_m = f(mh)$ and define Hankel and bordered Hankel matrices by

$$G_0[f_m] = 1,$$

$$G_k[f_m] = \begin{pmatrix} f_m & f_{m+1} & \cdots & f_{m+k-1} \\ f_{m+1} & f_{m+2} & \cdots & f_{m+k} \\ \vdots & \vdots & & \vdots \\ f_{m+k-1} & f_{m+k} & \cdots & f_{m+2k-2} \end{pmatrix}, \quad k = 1, 2, \ldots,$$

$$G_{k+1}[1; f_m] = \begin{pmatrix} 1 & 1 & \cdots & 1 \\ f_m & f_{m+1} & \cdots & f_{m+k} \\ f_{m+1} & f_{m+2} & \cdots & f_{m+k+1} \\ \vdots & \vdots & & \vdots \\ f_{m+k-1} & f_{m+k} & \cdots & f_{m+2k-1} \end{pmatrix}, \quad k = 0, 1, \ldots.$$

Denote their determinants as

$$H_k[f_m] = \det\{G_k[f_m]\},$$
$$H_{k+1}[1; f_m] = \det\{G_{k+1}[1; f_m]\}.$$

Finally we define

$$R_k(f_m) = \frac{H_k[f_m]}{H_k[1; f_m]},$$
$$S_k(f_m) = \frac{H_{k+1}[1; f_m]}{H_k[f_m]}. \quad \square$$

We call $\{S_k(f_m) \mid k = 0, 1, \ldots,\ m = \ldots, -1, 0, 1, \ldots,\}$ and $\{R_k(f_m) \mid k = 1, 2, \ldots,\ m = \ldots, -1, 0, 1, \ldots,\}$ the R array and the S array, respectively. Pye and Atchison (1973) presented a computationally efficient algorithm to calculate them recursively.

Algorithm 5.1. *The Pye-Atchison Algorithm*
If the $R_k(m)$ and $S_k(m)$ are defined, then they can be recursively calculated as follows:

$$S_0(f_m) = 1, \ m = 0, \pm 1, \pm 2, \ldots,$$

$$R_1(f_m) = f_m, \; m = 0, \pm 1, \pm 2, \ldots,$$
$$S_k(f_m) = S_{k-1}(f_{m+1}) \left(\frac{R_k(f_{m+1})}{R_k(f_m)} - 1 \right),$$
$$R_{k+1}(f_m) = R_k(f_{m+1}) \left(\frac{S_k(f_{m+1})}{S_k(f_m)} - 1 \right). \quad \square$$

Henceforth we denote $R_n(\rho_m)$ and $S_n(\rho_m)$ by $R_n(m)$ and $S_n(m)$, respectively. Property 5.1 determines the pattern of the S array.

Property 5.9. Let $\{y_t\}$ be from the ARMA(p, q) model (1.1). Then the following holds.

1. If $k = p, \; m = q - p + 1, q - p + 2, \ldots,$ then
$$S_p(m) = (-1)^{p+1} \sum_{j=0}^{p} \phi_j.$$
Also, $S_p(q - p) \neq (-1)^{p+1} \sum_{j=0}^{p} \phi_j$.
2. If $k = p, \; m = -q - p, -q - p - 1, \ldots,$ then
$$S_p(m) = \frac{(-1)^p}{\phi_p} \sum_{j=0}^{p} \phi_j.$$
Also, $S_p(-q - p + 1) \neq (-1)^p \sum_{j=0}^{p} \phi_j / \phi_p$.
3. If $k = p, p + 1, \ldots, \; m = q - k + 1$, then
$$S_k(q - k + 1) = (-1)^{k+1} \sum_{j=0}^{p} \phi_j.$$
4. If $k = p + 1, p + 2, \ldots, \; m = q - k$, then $S_k(q - k)$ is indeterminate. $\square$

The two-way array of the S pattern is tabulated in Table 5.4. The S pattern in Property 5.9 holds only for $p > 0$. Thus, it is not useful to identify a pure MA process. In this case, we should use the first column of the R array, which itself is the ACRF. Woodward and Gray (1981) have preferred the shifted S array, whose (m, k) element is $S_k(m + k - 1)$, to the S array.

Property 5.2 yields that the rank of $G_k(\rho_m)$ is

$$k - \min(k - p, m + k - 1 - q) \quad \text{if } m + k - 1 \geq q$$

$$k - \min(k - p, -m - k + 1 - q) \quad \text{if } m + k - 1 \leq -q.$$

Table 5.4. The S array

$m \setminus k$	$\cdots$	$p-1$	p	$p+1$	$p+2$	$\cdots$
$\vdots$		$\vdots$	$\vdots$	$\vdots$	$\vdots$	
$-q-p-2$	$\cdots$	$S_{p-1}(-q-p-2)$	C_2	∞	∞	$\cdots$
$-q-p-1$	$\cdots$	$S_{p-1}(-q-p-1)$	C_2	∞	NC	$\cdots$
$-q-p$	$\cdots$	$S_{p-1}(-q-p)$	C_2	NC	NC	$\cdots$
$-q-p+1$	$\cdots$	$S_{p-1}(-q-p+1)$	NC	NC	NC	$\cdots$
$\vdots$		$\vdots$	$\vdots$	$\vdots$	$\vdots$	
$q-p-1$	$\cdots$	$S_{p-1}(q-p-1)$	NC	NC	C_1	$\cdots$
$q-p$	$\cdots$	$S_{p-1}(q-p)$	NC	$-C_1$	∞	$\cdots$
$q-p+1$	$\cdots$	$S_{p-1}(q-p+1)$	C_1	∞	∞	$\cdots$
$q-p+2$	$\cdots$	$S_{p-1}(q-p+2)$	C_1	∞	∞	$\cdots$
$\vdots$		$\vdots$	$\vdots$	$\vdots$	$\vdots$	

Note: NC represents nonconstant.

Thus, $R_k(m)$ is not defined if (k, m) belongs to

$$\{(k,m) \mid k > p,\ m > q-k+1\} \bigcup \{(k,m) \mid k > p,\ m \leq -q-k\}.$$

Therefore, the R array in Definition 5.1 has no useful pattern for ARMA model identification. However, it can be modified so that it has an appropriate pattern.

Theorem 5.7. Let $\{\rho_k\}$ be any sequence such that $B(k, m+k-1)$ and $B(k-1, m+k-2)$ are nonsingular. If $m+k-2 \geq 0$, then

$$R_k(m) = (-1)^k \frac{\theta(k-1, m+k-2)}{\sum_{j=0}^{k-1} \phi_{k-1,j}^{(m+k-2)}}.$$

If $m+k-2 < 0$, then

$$R_k(m) = (-1)^k \frac{\eta(k-1, -m-k+2)}{\sum_{j=0}^{k-1} \pi_{k-1,j}^{(-m-k+2)}}. \quad \square$$

Henceforth we regard $R_k(m)$ in Theorem 5.7 as the definition of the (k, m) element of the R array as long as $B(k-1, m+k-2)$ is nonsingular. The EYW equations imply the following property.

Property 5.10 Let $\{y_t\}$ be from the ARMA(p, q) model (1.1). Then the following holds.

1. If $k = p+1$, $m = q-p+1, q-p+2, \ldots$, then $R_{p+1}(m) = 0$. Also, $R_{p+1}(q-p) \neq 0$.

Table 5.5. The R array

$m \setminus k$	$\cdots$	$p-1$	p	$p+1$	$p+2$	$\cdots$
$\vdots$		$\vdots$	$\vdots$	$\vdots$	$\vdots$	
$-q-p-1$	$\cdots$	$R_{p-1}(-q-p-1)$	0	0	X	$\cdots$
$-q-p$	$\cdots$	$R_{p-1}(-q-p)$	0	X	X	$\cdots$
$-q-p+1$	$\cdots$	$R_{p-1}(-q-p+1)$	X	X	X	$\cdots$
$\vdots$		$\vdots$	$\vdots$	$\vdots$	$\vdots$	
$q-p-1$	$\cdots$	$R_{p-1}(q-p-1)$	X	X	0	$\cdots$
$q-p$	$\cdots$	$R_{p-1}(q-p)$	X	0	0	$\cdots$
$q-p+1$	$\cdots$	$R_{p-1}(q-p+1)$	0	0	0	$\cdots$
$q-p+2$	$\cdots$	$R_{p-1}(q-p+2)$	0	0	0	$\cdots$
$\vdots$		$\vdots$	$\vdots$	$\vdots$	$\vdots$	

Note: X represents nonzero.

2. If $k = p+1, p+2, \ldots,\ m = q+2-k$, then $R_k(q+2-k) = 0$.
3. If $k = p+1,\ m = -p-q-1, -p-q-2, \ldots,$ then $R_{p+1}(m) = 0$. Also, $R_{p+1}(-q-p) \neq 0$. □

The two-way array of the R pattern is tabulated in Table 5.5. We can represent the S function using $\{\phi_{k,j}^{(i)}\}$ and $\{\pi_{k,j}^{(i)}\}$.

Property 5.11. If $B(k, m+k-1)$ is nonsingular, then

$$\begin{aligned} S_k(m) &= (-1)^{k+1} \sum_{j=0}^{k} \phi_{k,j}^{(m+k-1)} \\ &= (-1)^{k+1} \sum_{j=0}^{k} \pi_{k,j}^{(-m-k+1)}. \end{aligned}$$

Moreover, if $B(k, -m-k)$ is also nonsingular, then

$$S_k(m) = \frac{(-1)^k}{\phi_{k,k}^{(-m-k)}} \sum_{j=0}^{k} \phi_{k,j}^{(-m-k)}. \quad \square$$

The R and S array method is closely related to the Padé approximation (see, e.g., Baker [1975]), Aitken's Δ^2 method (see, e.g., Conte and Boor [1987, pp. 95-99]), and the jackknife statistic (see, e.g., Gray and Schucany [1972]). As pointed out by O. D. Anderson (1980a), some statisticians are enthusiastic about the R and S array method, while others are less optimistic. As Gooijer and Heuts (1981) have mentioned, it is, above all, its complexity and theoretical difficulty that restrains us from utilizing the R and S array method.

5.2.2 Asymptotic Distributions

Choi (1990f) presented the asymptotic distributions of the R and the S arrays.

Theorem 5.8. *The Asymptotic Distribution of the S Array*
Let $\{y_1, \ldots, y_T\}$ be a T-realization from the ARMA(p, q) model (1.1). Then the following holds.

1. If $k = p$ and $m, n = q-p+1, q-p+2, \ldots$, then $\sqrt{T}\left\{\hat{S}_p(m) - S_p(m)\right\}$ and $\sqrt{T}\left\{\hat{S}_p(n) - S_p(n)\right\}$ are asymptotically normally distributed with means 0 and covariance
$$\sigma^2 \mathbf{1}^t \Sigma^{-1}(p, m+p-1) \sum_{\alpha=0}^{q} \sum_{\beta=0}^{q} \theta_\alpha \theta_\beta \Sigma(p, m-n+\beta-\alpha) \Sigma^{-1}(p, n+p-1)^t \mathbf{1},$$
where $\mathbf{1} = (1, \ldots, 1)^t$.

2. If $k = p$ and $m, n = -q-p, -q-p-1, \ldots$, then $\sqrt{T}\left\{\hat{S}_p(m) - S_p(m)\right\}$ and $\sqrt{T}\left\{\hat{S}_p(n) - S_p(n)\right\}$ are asymptotically normally distributed with means 0 and covariance
$$\frac{\sigma^2}{\phi_p^2} \boldsymbol{\alpha}^t \Sigma^{-1}(p, -m-p) \sum_{\alpha=0}^{q} \sum_{\beta=0}^{q} \theta_\alpha \theta_\beta \Sigma(p, n-m+\beta-\alpha) \Sigma^{-1}(p, -n-p)^t \boldsymbol{\alpha},$$
where
$$\boldsymbol{\alpha} = \left(1, \ldots, 1, 1 - \frac{1}{\phi_p} \sum_{j=0}^{p} \phi_j\right)^t.$$

3. If $k, j = p, p+1, \ldots$, then $\sqrt{T}\left\{\hat{S}_k(q-k+1) - S_k(q-k+1)\right\}$ and $\sqrt{T}\left\{\hat{S}_j(q-j+1) - S_j(q-j+1)\right\}$ are asymptotically normally distributed with means 0 and covariance
$$(-1)^{k+j} \sigma^2 \mathbf{1}^t \Sigma^{-1}(k, q) \sum_{\alpha=0}^{q} \sum_{\beta=0}^{q} \theta_\alpha \theta_\beta \Sigma(k, j; \beta - \alpha) \Sigma^{-1}(j, q)^t \mathbf{1},$$
where $\Sigma(k, j; \beta - \alpha)$ is the $k \times j$ matrix whose (r, s) element is $\sigma(r - s + \beta - \alpha)$. □

Theorem 5.9. *The Asymptotic Distribution of the R Array*
Let $\{y_1, \ldots, y_T\}$ be a T-realization from the ARMA(p, q) model (1.1). Then the following holds.

1. If $k = p+1$ and $m, n = q-p+1, q-p+2, \ldots$, then $\sqrt{T}\left\{\hat{R}_{p+1}(m) - R_{p+1}(m)\right\}$ and $\sqrt{T}\left\{\hat{R}_{p+1}(n) - R_{p+1}(n)\right\}$ are asymptotically normally distributed with means 0 and covariance

$$\frac{\sigma^2}{\sigma(0)(\sum_{j=0}^{p}\phi_j)^2}\sum_{r=0}^{p}\sum_{s=0}^{q}\sum_{\alpha=0}^{p}\sum_{\beta=0}^{q}\phi_r\theta_s\phi_\alpha\theta_\beta\rho_{m-n-r-s+\alpha+\beta}.$$

2. If $k, j = p+1, p+2, \ldots$, then $\sqrt{T}\left\{\hat{R}_k(q-k+2) - R_k(q-k+2)\right\}$ and $\sqrt{T}\left\{\hat{R}_j(q-j+2) - R_j(q-j+2)\right\}$ are asymptotically normally distributed with means 0 and covariance

$$(-1)^{k+j}\frac{\sigma^2}{\sigma(0)(\sum_{j=0}^{p}\phi_j)^2}\sum_{r=0}^{p}\sum_{s=0}^{q}\sum_{\alpha=0}^{p}\sum_{\beta=0}^{q}\phi_r\theta_s\phi_\alpha\theta_\beta\rho_{k-j-r-s+\alpha+\beta}. \quad \square$$

The identification procedure using the R and S arrays is a pattern recognition method. GKM suggested the D statistic to measure agreement with the proper pattern for a stationary ARMA process. (Readers should note that it is different from the D array of the Corner method.) GKM concluded that it performs much better than the AIC procedure with less computing time. However, Tucker (1982) presented some examples where the D statistic results in incorrect choices of the orders. For the same purpose as the D statistic, we can construct some χ^2 statistics based on the asymptotic distributions of Theorems 5.8 and 5.9. There are more references about the R and S array method in Section 5.9.

5.2.3 The *RS* Array

Theorem 5.7 and Property 5.11 imply the following.

Corollary 5.2. If the corresponding R and S terms are defined, then

$$\theta(k,i) = R_{k+1}(i-k+1)S_k(i-k+1),$$

$$\lambda(k,i) = -R_{k+1}(i-k+1)S_k(-k-1-i)\frac{S_k(i-k+1)}{S_{k+1}(i-k)}. \quad \square$$

Corollary 5.2 shows that the RS array is equivalent to the θ array.

Property 5.12. Let $\{y_t\}$ be from the ARMA(p,q) model (1.1). Then, the following holds.

1. $R_{k+1}(i)S_k(i) = 0$, $k = p$ and ($i \geq q-p+1$ or $i \leq -q-p$).

Table 5.6. The Corner array

$k \backslash i$	$\cdots$	q	$q+1$	$q+2$	$q+3$	$\cdots$
$\vdots$		$\vdots$	$\vdots$	$\vdots$	$\vdots$	
p	$\cdots$	X	X	X	X	$\cdots$
$p+1$	$\cdots$	X	0	0	0	$\cdots$
$p+2$	$\cdots$	X	0	0	0	$\cdots$
$\vdots$		$\vdots$	$\vdots$	$\vdots$	$\vdots$	

Note: X represents nonzero.

2. $R_{k+1}(i)S_k(i) = 0$, $k \geq p$ and $(i = q - p + 1$ or $i = -q - p)$. □

Tucker (1982) proposed to use the RS array instead of the R and the S arrays in ARMA model identification.

5.3 The Corner Method

5.3.1 Correlation Determinants

Let $D(k,i)$ be the determinant of $B(k,i)$. We call $\{D(k,i) \mid k = 1,2,\ldots, i = 0,1,\ldots\}$ the D array of the ARMA process. The D array is equivalent to the C table that appears in the Padé approximation. For more details, readers may consult the references in Section 5.9. Property 5.1 implies the D array has the following pattern.

Property 5.13. A stochastic process has the ARMA(p,q) representation (1.1) *if and only if*

1. $D(k,i) = 0, \ (k,i) \in J_{p+1,q+1}$,
2. $D(k,i) \neq 0, \ (k,i) \in I_{p,q}$. □

The two-way array of the D pattern is shown in Table 5.6. The D array identification method was proposed by Beguin, Gourieroux, and Monfort (1980) (BGM). Some, like Rezayat and Anandalingam (1988), favor the Corner method. However, others, like Gooijer and Heuts (1981), do not recommend it for practical applications. The Corner method has been applied to identification of transfer function models by Liu and Hanssens (1982) and Lii (1985).

The D array can be represented by the three functions.

Property 5.14. If $B(k,i)$ is nonsingular, then the following holds.

1. $D(k+1,i+1) = (-1)^k \theta(k,i) D(k,i)$.

2. $D(k+1, i) = \lambda(k, i)D(k, i)$.
3. $D(k+1, i-1) = (-1)^k \eta(k, i)D(k, i)$. $\square$

Property 5.14 implies that the D array can be calculated through the simplified Trench-Zohar algorithm. Pham (1984b) presented the relation

$$D^2(k, i) = D(k, i+1)D(k, i-1) + D(k+1, i)D(k-1, i)$$

and suggested using it to compute the D array. We can easily derive Pham's relation using Property 5.14.

5.3.2 Asymptotic Distribution

One of the reasons for not using the D array is the difficulty of simultaneous testing of the zeroness. To test the zero behavior of the lower-right corner of the two-way D array, we need the joint distribution of the D array estimate. BGM proposed a statistic G_n to test the null hypothesis,

$$H_0: \ D(k, i) = 0, \quad (k, i) \in J_{p+1,q+1}.$$

Let $\{D_1, \cdots, D_n\}$ be a subset of $\{D(k, i) \mid (k, i) \in J_{p+1,q+1}\}$. Then, a test statistic is defined as

$$G_n = T(\hat{D}_1, \ldots, \hat{D}_n)[\hat{A}\hat{G}_H\hat{A}^t]^{-1}(\hat{D}_1, \ldots, \hat{D}_n)^t,$$

where the (i, j) element of A is $\partial D_i / \partial \rho_j$, H is the maximal index h of the ρ_h's appearing in $D_1, \ldots, D_n$, and the (i, j) element of G_H is $\text{cov}(\hat{\rho}_i, \hat{\rho}_j)$, which is presented in Theorem 1.2. Under the null hypothesis H_0, G_n converges in distribution to the χ_n^2 random variable. The asymptotic distribution is useful only when the covariance matrix $\hat{A}\hat{G}_H\hat{A}^t$ is nonsingular. Gooijer and Heuts (1981) and Petruccelli and Davies (1984) have shown that the nonsingularity condition imposes some restrictions on the use of the G_n statistic.

Choi (1990e) presented a closed form of the asymptotic joint distribution of the estimate of the D array.

Theorem 5.10. *The Asymptotic Distribution of the D Array*
Let $\{y_1, \ldots, y_T\}$ be a T-realization from the ARMA(p, q) model (1.1). Then the following holds.

1. If $k, j = p, p+1, \ldots$, then $\sqrt{T}\hat{D}(k+1, q+1)$ and $\sqrt{T}\hat{D}(j+1, q+1)$ are asymptotically normally distributed with means 0 and covariance

$$\begin{aligned} a_{k,j} &= \lim_{T\to\infty} \text{cov}\left(\sqrt{T}\hat{D}(k+1, q+1), \sqrt{T}\hat{D}(j+1, q+1)\right) \\ &= (-1)^{k+j} D(k, q)D(j, q)\frac{1}{\sigma^2(0)} \sum_{m=-q}^{q} \sigma_z(m)\sigma_z(m+k-j), \end{aligned}$$

where

$$\sigma_z(m) = \sum_{r=0}^{p} \sum_{s=0}^{p} \phi_r \phi_s \sigma(m + r - s).$$

2. If $i, j = q, q+1, \ldots$, then $\sqrt{T}\hat{D}(p+1, i+1)$ and $\sqrt{T}\hat{D}(p+1, j+1)$ are asymptotically normally distributed with means 0 and covariance

$$\begin{aligned} b_{i,j} &= \lim_{T\to\infty} \operatorname{cov}\left(\sqrt{T}\hat{D}(p+1,i+1), \sqrt{T}\hat{D}(p+1,j+1)\right) \\ &= D(p,i+1)D(p,j+1)\frac{1}{\sigma^2(0)\phi_p^2} \sum_{m=-q}^{q} \sigma_z(m)\sigma_z(m+i-j). \end{aligned}$$

3. If $k = p, p+1, \ldots, i = q, q+1, \ldots$, then $\sqrt{T}\hat{D}(k+1, q+1)$ and $\sqrt{T}\hat{D}(p+1, i+1)$ are asymptotically normally distributed with means 0 and covariance

$$\begin{aligned} c_{k,i} &= \lim_{T\to\infty} \operatorname{cov}\left(\sqrt{T}\hat{D}(k+1,q+1), \sqrt{T}\hat{D}(p+1,i+1)\right) \\ &= (-1)^{k-1} D(k,q)D(p,i+1)\frac{1}{\sigma^2(0)\phi_p} \\ &\quad \times \sum_{m=-q}^{q} \sigma_z(m)\sigma_z(m+k-p-i+q). \quad \square \end{aligned}$$

From Theorem 5.10 we can easily see the restrictions of the G_n statistic. Using the asymptotic distribution Choi (1990e) has made some χ^2 statistics to test H_0, which do not have the restrictions the G_n statistic does.

5.4 The GPAC Methods

5.4.1 Woodward and Gray's GPAC

The coefficient $\phi_{k,k}^{(i)}$ was named the generalized partial autocorrelation by Woodward and Gray (1981) (WG). We call $\{\phi_{k,k}^{(i)}\}$ WG's generalized partial autocorrelation function (GPAC). Property 5.1 implies the following property.

Property 5.15. Let $\{y_t\}$ be from the ARMA(p,q) model (1.1) with $p > 0$. Then, the following holds.

1. $\phi_{p,p}^{(i)} = \phi_p, \quad i = q, q+1, \ldots.$
2. $\phi_{k,k}^{(q)} = 0, \quad k = p+1, p+2, \ldots. \quad \square$

Table 5.7. Woodward and Gray's GPAC array

$k \setminus i$	$\cdots$	$q-1$	q	$q+1$	$q+2$	$\cdots$
$\vdots$		$\vdots$	$\vdots$	$\vdots$	$\vdots$	
p	$\cdots$	$\phi_{p,p}^{(q-1)}$	ϕ_p	ϕ_p	ϕ_p	$\cdots$
$p+1$	$\cdots$	$\phi_{p+1,p+1}^{(q-1)}$	0	U	U	$\cdots$
$p+2$	$\cdots$	$\phi_{p+2,p+2}^{(q-1)}$	0	U	U	$\cdots$
$p+3$	$\cdots$	$\phi_{p+3,p+3}^{(q-1)}$	0	U	U	$\cdots$
$\vdots$		$\vdots$	$\vdots$	$\vdots$	$\vdots$	

Note: U represents undefined.

The two-way array of WG's GPAC is tabulated in Table 5.7. Readers should note that the roles of rows and columns in the table are reversed from those of WG's paper.

The idea of using the GPAC in ARMA model identification originated from Bartlett and Diananda (1950, Section 5). The GPAC can be represented by a ratio of the three functions or that of the S function.

Property 5.16. The GPAC can be represented by

$$\phi_{k,k}^{(i)} = \frac{\theta(k-1,i)}{\lambda(k-1,i)},$$

$$\phi_{k,k}^{(i)} = \frac{S_k(-k+i+1)}{S_k(-k+i)}. \quad \square$$

Property 5.16 implies some significant information of the three functions and that of the S array may be lost when the ratio is taken. Because of the lost information, WG have emphasized that the patterns in the GPAC array should be always checked via the S array.

If the underlying process is from the ARMA(p,q) model (1.1), the constant behavior of $\hat{\phi}_{p,p}^{(i)}$ will not be kept well as i increases, for the variance of $\hat{\phi}_{p,p}^{(i)}$ increases as i does. WG have mentioned that the GPAC method may identify the underlying process as coming from an ARMA(p,j) model where $j < q$. Thus, we should be cautious when we use the GPAC array for determining q. Davies and Petruccelli (1984) have presented a paper arguing against the use of the GPAC in ARMA model identification. They have claimed that the GPAC shows an unstable behavior when it is applied to time series of moderate length. Moreover, its use in detecting the orders of MA parts in real series is very limited, and thus it can only be recommended as a means of confirming a pure AR fit to the data. To demonstrate difficulties in the application of the GPAC technique, they provided simulations and real data analyses and made the following conclusions.

- A positive identification can rarely be made for data that are well established to be mixed ARMA.
- When a positive identification can be made, it seems to agree only with previous analyses for pure AR or near pure AR identification.
- The presence of MA terms in the model to be identified appears to alter dramatically the viability of the GPAC array technique even for moderate sample sizes.

Newbold and Bos (1983) have shown in simulation studies for ARMA(1, 1) processes that the empirical variances of $\hat{\phi}_{1,1}^{(i)}$, $i = 2, 3, \ldots$, are large even for 100 observations, and then have concluded that the GPAC estimator has a heavy tailed distribution. Davies and Petruccelli (1984, p. 374) have pointed out that the numerator $\hat{\theta}(k-1, i)$ of $\hat{\phi}_{k,k}^{(i)}$ has well-behaved asymptotic properties and satisfactory sampling properties in large samples. However, the nice property of the asymptotic distribution of $\hat{\theta}(k-1, i)$ cannot be applicable to the estimate of the GPAC because $\hat{\lambda}(k-1, i)$ is small in absolute value.

Choi (1991a) derived the asymptotic distributions of WG's GPAC.

Theorem 5.11. Let $\{y_1, \ldots, y_T\}$ be a T-realization from the ARMA(p, q) model (1.1). If $k, j = p+1, p+2, \ldots$, then $T^{1/2}\hat{\phi}_{k,k}^{(q)}$ and $T^{1/2}\hat{\phi}_{j,j}^{(q)}$ are asymptotically normally distributed with means 0 and covariance

$$\frac{1}{\sigma^2(0)\lambda^2(p,q)} \sum_{m=-q}^{q} \sigma_z(m)\sigma_z(m+k-j),$$

where

$$\sigma_z(m) = \sum_{r=0}^{p} \sum_{s=0}^{p} \phi_r \phi_s \sigma(m+r-s). \quad \square$$

Theorem 5.12. Let $\{y_1, \ldots, y_T\}$ be a T-realization from the ARMA(p, q) model (1.1). If $i = p+1, p+2, \ldots$, then, as $T \to \infty$, $\hat{\phi}_{p+1,p+1}^{(i)}$ converges in distribution to

$$\frac{\{a^2(0) - a^2(1)\}^{1/2}\phi_p}{a(0)} W - \frac{a(1)}{a(0)}\phi_p,$$

where W is the standard Cauchy random variable and

$$a(0) = \frac{1}{\sigma^2(0)} \sum_{j=-q}^{q} \sigma_z^2(j),$$

Table 5.8. Glasbey's GPAC array

$k \setminus i$	$\cdots$	$q-1$	q	$q+1$	$\cdots$
$\vdots$		$\vdots$	$\vdots$	$\vdots$	
$p-1$	$\cdots$	$\zeta(p-1,q-1)$	$\zeta(p-1,q)$	$\zeta(p-1,q+1)$	$\cdots$
p	$\cdots$	$\zeta(p,q-1)$	0	0	$\cdots$
$p+1$	$\cdots$	$\zeta(p+1,q-1)$	0	$\zeta(p+1,q+1)$	$\cdots$
$p+2$	$\cdots$	$\zeta(p+2,q-1)$	0	$\zeta(p+2,q+1)$	$\cdots$
$\vdots$		$\vdots$	$\vdots$	$\vdots$	

$$a(1) = \frac{1}{\sigma^2(0)} \sum_{j=-q}^{q} \sigma_z(j)\sigma_z(j+1). \quad \Box$$

Based on the asymptotic distributions and some computational experience, the present author understands that the denominator $\lambda(p-1,i)$ is very small in absolute value for $i = q+1, q+2, \ldots$ and becomes smaller as i increases. So the variance of the GPAC estimator should be large. There are more references about WG's GPAC in Section 5.9.

5.4.2 Glasbey's GPAC

Glasbey (1982) presented the following statistic for ARMA model identification:

$$\zeta(k,i) = \frac{\rho_{i+k+1} - \sum_{j=1}^{k} \phi_{k,j}^{(i)} \rho_{i+k+1-j}}{\sqrt{\sum_{j=-i}^{i} v_j^2(k,i)}},$$

where

$$v_j(k,i) = \sum_{m=0}^{\min(k,i-j)} \sum_{l=0}^{k} \phi_{k,m}^{(i)} \phi_{k,l}^{(i)} \rho_{j+m-l}.$$

It can be shown that $\zeta(k,0) = \phi_{k,k}^{(0)}$. Thus, $\{\zeta(k,i)\}$ can be considered as a generalization of the PACF. We will call it Glasbey's GPAC. Because the numerator of $\zeta(k,i)$ is $\theta(k,i)$ and the denominator is positive, Property 5.3 implies that Glasbey's GPAC has the following pattern.

Property 5.17. Let $\{y_t\}$ be from the ARMA(p,q) model (1.1). Then

$$\zeta(k,i) = 0, \quad (k,i) \in I_{p,q}. \quad \Box$$

The two-way array of Glasbey's GPAC is tabulated in Table 5.8. Glasbey (1982, p. 224) has mentioned without proof that if either $k < p$ or $i < q$, then $\zeta(k,i) \neq 0$. But this is still an open question.

Glasbey has derived the asymptotic distribution of $\{\hat{\zeta}(k,i)\}$ using the technique due to Quenouille (1947). However, we can derive it from Theorem 5.1.

Theorem 5.13. Let $\{y_1,\dots,y_T\}$ be a T-realization from the ARMA(p,q) model (1.1). If $(k,i)\in I_{p,q}$, then the $\sqrt{T}\hat{\zeta}(k,i)$'s are asymptotically normally distributed with means 0 and variances 1. Additionally, if $(j,l)\in I_{p,q}$, then the asymptotic correlation of $\sqrt{T}\hat{\zeta}(k,i)$ and $\sqrt{T}\hat{\zeta}(j,l)$ is

$$\frac{\sum_{m=-q}^{q} v_m(p,q)v_{m+k+i-j-l}(p,q)}{\sum_{m=-q}^{q} v_m^2(p,q)}. \quad \square$$

Pham (1984b) has derived a recursive formula of some random variables to which $\sqrt{T}\hat{\zeta}(k,i)$'s converge in distribution for $k=p,p+1,\dots,$ $i=q,q+1,\dots.$

Based on simulation results, Glasbey has concluded that the ζ array is a very good tool for identifying ARMA processes if T is large and that there is a tendency to underestimate the orders if T is small.

5.4.3 Takemura's GPAC

Takemura (1984) proposed a generalization of both the ACRF and the PACF as follows. If $B(k,i)$ is nonsingular, then

$$\tau(k,i)=\frac{\rho_{i+k+1}-\sum_{j=1}^{k}\phi_{k,j}^{(i)}\rho_{i+k+1-j}}{1-2\boldsymbol{\rho}(k,0)^t\boldsymbol{\phi}(k,i)+\boldsymbol{\phi}(k,i)^tB(k,0)\boldsymbol{\phi}(k,i)}.$$

If $B(k,i)$ is singular, then $\tau(k,i)=0$. We call $\{\tau(k,i)\}$ Takemura's GPAC.

Property 5.18. If the underlying process is from the ARMA(p,q) model (1.1), then the following holds.

1. $\tau(0,i)=\rho_i,\ i=0,1,\dots.$
2. $\tau(k,0)=\phi_{k,k}^{(0)},\ k=0,1,\dots.$
3. $|\tau(k,i)|\le 1$ for any (k,i). $\square$

Property 5.18 shows that $\{\tau(k,i)\}$ is a generalization of the ACRF and the PACF and that $\tau(k,i)$ is a correlation coefficient of certain random variables. Because the numerator of $\tau(k,i)$ is $\theta(k,i)$, Property 5.3 implies the following property.

Property 5.19. If the underlying process is from the ARMA(p,q) model (1.1), then

$$\tau(k,i)=0,\quad (k,i)\in J_{p,q}. \quad \square$$

Table 5.9. Takemura's GPAC array

$k \setminus i$	0	$\cdots$	$q-1$	q	$q+1$	$q+2$	$\cdots$
0	1	$\cdot$	ρ_{q-1}	ρ_q	ρ_{q+1}	ρ_{q+2}	$\cdots$
$\vdots$	$\vdots$		$\vdots$	$\vdots$	$\vdots$	$\vdots$	
p	$\phi_{p,p}^{(0)}$	$\cdots$	$\tau(p, q-1)$	0	0	0	$\cdots$
$p+1$	$\phi_{p+1,p+1}^{(0)}$	$\cdots$	$\tau(p+1, q-1)$	0	0	0	$\cdots$
$p+2$	$\phi_{p+2,p+2}^{(0)}$	$\cdots$	$\tau(p+2, q-1)$	0	0	0	$\cdots$
$\vdots$	$\vdots$		$\vdots$	$\vdots$	$\vdots$	$\vdots$	

The two-way table of Takemura's GPAC is tabulated in Table 5.9.

Takemura has derived the asymptotic distribution of $\{\hat{\tau}(k,i)\}$ using Bartlett and Diananda's method (1950). However, we can derive it using Theorem 5.1.

Theorem 5.14. Let $\{y_1, \ldots, y_T\}$ be a T-realization from the ARMA(p,q) model (1.1). If $(k,i) \in I_{p,q}$, then $\sqrt{T}\hat{\tau}(k,i)$'s are asymptotically normally distributed with means 0. Additionally, if $(j,l) \in I_{p,q}$, then the asymptotic covariance of $\sqrt{T}\hat{\tau}(k,i)$ and $\sqrt{T}\hat{\tau}(j,l)$ is

$$\sum_{m=-q}^{q} \rho_m(z)\rho_{m-k-i+j+l}(z),$$

where $\{\rho_m(z)\}$ is the ACRF of $z_t = \phi(B)y_t = \theta(B)v_t$. $\square$

Takemura proposed a recursive algorithm to calculate $\{\hat{\tau}(k,i)\}$, which is similar to the simplified Trench-Zohar algorithm in Section 1.3.

5.5 The ESACF Method

TT-84 presented a unified procedure to identify both stationary and nonstationary ARMA processes. They used the ILS method discussed in Section 1.5 to produce consistent estimates of the AR parameters, and then defined the extended sample autocorrelation function (ESACF) based on the estimates. This procedure has a clear advantage in that it eliminates the need to determine the order of differencing necessary to produce stationarity in ARMA modeling. However, in order to obtain the (k,i) element of the ESACF array, it is necessary to calculate the ILS estimates of the AR parameters of the ARMA(k,i) model, the residuals of the regression, and their sample autocorrelations. Thus, this procedure requires more calculations than any other pattern identification method.

TT-84 and Tiao (1985) have proposed the ESACF $\{r_{i(k)}\}$ for ARMA model identification as follows.

1. For $k = 0$, let $r_{i(0)} = \hat{\rho}_i$.

2. For $k = 1, 2, \ldots$, let

$$w_{k,t}^{(i)} = y_t - \sum_{l=1}^{k} \hat{\phi}_{l(k)}^{(i)} y_{t-l}.$$

   The extended sample autocorrelation $r_{i(k)}$ is defined as the sample autocorrelation at lag i of $\{w_{k,t}^{(i)}\}$.

If the true orders are p and q, then $\{w_{p,t}^{(i)}\}$ follows an MA(q) model for $i = q, q+1, \ldots$. Therefore, the ESACF satisfies the following.

Property 5.20. *The ESACF Pattern*
Let $\{y_1, \ldots, y_T\}$ be a T-realization from the ARMA(p, q) model (1.3). Then the following holds.

1. $r_{q(p)} \neq 0$.

2. $r_{i(p)} \overset{P}{=} 0, \;\; i = q+1, q+2, \ldots$, where $\overset{P}{=}$ means asymptotic equivalence in probability. □

Property 5.20 is useful for identifying the MA order q when the AR order p is known. When both of the orders are unknown, a generalized property should be applied.

Property 5.21. *The Generalized ESACF Pattern*
Let $\{y_1, \ldots, y_T\}$ be a T-realization from the ARMA(p, q) model (1.3). Then the following holds.

1. $r_{i(k)} \overset{P}{=} 0, \;\; i - q > k - p \geq 0$.

2. $r_{i(k)} \overset{P}{=} c(k-p, i-q), \;\; k - p \geq i - q \geq 0$, where $c(k-p, i-q)$ is a nonzero constant or a continuous random variable bounded between -1 and 1. □

The two-way array of the generalized ESACF is tabulated in Table 5.10. It should be noted that the pattern of $\{(k, i) \mid k - p > i - q \geq 0\}$ is not clear.

Table 5.10. The ESACF array

$k \setminus i$	$\cdots$	q	$q+1$	$q+2$	$\cdots$
$\vdots$		$\vdots$	$\vdots$	$\vdots$	
$p-1$	$\cdots$	X	X	X	$\cdots$
p	$\cdots$	X	0	0	$\cdots$
$p+1$	$\cdots$	X	X	0	$\cdots$
$p+2$	$\cdots$	X	X	X	$\cdots$
$\vdots$		$\vdots$	$\vdots$	$\vdots$	

Note: X represents nonzero value.

If the underlying process is stationary, then $r_{i(k)}$ is an estimate of the correlation of $\sum_{r=0}^{p} \phi_r y_{t-r}$ and $\sum_{s=0}^{p} \phi_s y_{t-i-s}$, which is

$$\sum_{r=0}^{p} \sum_{s=0}^{p} \phi_r \phi_s \rho_{i+s-r}.$$

If $i \geq q$, then it equals $\lambda(p, i)$.

In order to use the ESACF pattern for ARMA model identification, we need the asymptotic distribution of $\{r_{i(k)}\}$. TT-84 used the value $1/(T-k-i)$ as its sample variance, which is due to Theorem 1.2. However, numerical examples show that some elements of the ESACF array that are in the triangle of zeros are larger than two times the crude standard deviation in absolute value.

TT-84 have suggested the possibility of using another form of the ESACF, $\{r^*_{i(k)}\}$, which is defined as the sample ACRF of

$$w_{k,t}^{(i-1)} = y_t - \sum_{l=1}^{k} \hat{\phi}_{l(k)}^{(i-1)} y_{t-l}.$$

It has the following pattern.

Property 5.22. Let $\{y_1, \ldots, y_T\}$ be a T-realization from the ARMA(p, q) model (1.3), then

$$r^*_{i(k)} \stackrel{P}{=} 0, \quad (k, i) \in J_{p,q}. \quad \square$$

Based on their experiences, TT-84 have concluded that $\{r_{i(k)}\}$ is a better tool than $\{r^*_{i(k)}\}$. The present author guesses it would be easier to derive the asymptotic joint distribution of the latter than that of the former.

Finally it should be mentioned that we can apply the ILS estimates instead of the EYW estimates to any pattern identification method. This will be particularly useful when some roots of $\phi(z) = 0$ are near or on the unit circle. There are more references about the ESACF method in Section 5.9.

5.6 The SCAN Method

In order to determine the orders of stationary or nonstationary univariate ARMA processes, TT-85 proposed the smallest canonical correlation (SCAN) method. Before studying the SCAN method, we consider eigen-analysis of the correlation matrix of an ARMA process.

5.6.1 EIGEN-ANALYSIS

Because the determinant of a matrix is the product of its eigenvalues, Property 5.2 implies the following.

Property 5.23. Let $\nu(k,i)$ be the smallest eigenvalue of $B(k,i)$ in absolute value. Then, a stochastic process has the ARMA(p,q) representation (1.1) *if and only if*

1. $\nu(k,i) = 0, \ (k,i) \in J_{p+1,q+1}$,
2. $\nu(k,i) \neq 0, \ (k,i) \in I_{p,q}$. □

If $k = p+1, p+2, \ldots,\ i = q+1, q+2, \ldots$, then the EYW equations imply

$$\Sigma(k,i)\boldsymbol{\phi}_*(k) = \mathbf{0},$$

where $\boldsymbol{\phi}_*(k) = (\phi_0, \ldots, \phi_p, 0, \ldots, 0)^t$. Thus, $\boldsymbol{\phi}_*(k)$ is an eigenvector of $\Sigma(k,i)$ corresponding to the eigenvalue 0, and the eigen-analysis provides information not only about the orders but also about the parameters. These properties might be used for ARMA modeling. However, this method has no advantage over any other pattern identification method. In particular it is computationally much more expensive than any other method, for no recursive algorithm for nested nonsymmetric Toeplitz matrices has been presented yet. Moreover, it will not be easy to find the asymptotic distribution of $\hat{\nu}(k,i)$.

Consider the matrix $B_2(k,i) = B(k,i)^t B(k,i)$. Denote its smallest eigenvalue by $\nu_2(k,i)$. Then, $\nu_2(k,i)$ satisfies the following property.

Property 5.24. A stochastic process has the ARMA(p,q) representation (1.1) *if and only if*

1. $\nu_2(k,i) = 0, \ (k,i) \in J_{p+1,q+1}$,
2. $\nu_2(k,i) \neq 0, \ (k,i) \in I_{p,q}$. □

Because $B_2(k,i)$ is non-negative definite, we may find some iterative procedures for computing the eigenvalues and the eigenvectors. Also, we may find the asymptotic distribution of $\hat{\nu}_2(k,i)$ using that of the eigenvalue of a Wishart matrix (see, e.g., T. W. Anderson [1984, p. 532]). This eigen-analysis is still more costly than any other pattern identification method. If $k = p+1, p+2, \ldots,\ i = q+1, q+2, \ldots,$ then $\boldsymbol{\phi}_*(k)$ is an eigenvector corresponding to the eigenvalue 0.

Consider another modification of $\nu(k,i)$. Let $\nu_*(k,i)$ be the smallest eigenvalue of $B_*(k,i) = \Sigma(k,i)^t\Sigma^{-1}(k,0)\Sigma(k,i)$. Akaike (1974b, 1975, 1976) has proposed the use of $\nu_*(k,i)$, which is the correlation between the future and the past variables. Piccolo and Tunnicliffe-Wilson (1984) have utilized the covariance matrix to explain some identification methods like the Corner method, the R and S array method, Woodward and Gray's GPAC, Glasbey's GPAC, the SCAN method which will be discussed in the next subsection, the HR method, and so on. Because $\Sigma(k,0)$ is positive definite, Property 5.23 implies the following.

Property 5.25. A stochastic process has the ARMA(p,q) representation (1.1) *if and only if*

1. $\nu_*(k,i) = 0,\ (k,i) \in J_{p+1,q+1},$
2. $\nu_*(k,i) \neq 0,\ (k,i) \in I_{p,q}.$ □

Because $B_*(k,i)$ is positive definite, we may find some iterative procedure for computing its eigenvalues and eigenvectors. Also, we may find the asymptotic distribution of $\hat{\nu}_*(k,i)$. If $k = p+1, p+2, \ldots, i = q+1, q+2, \ldots,$ then $\boldsymbol{\phi}_*(k)$ is an eigenvector corresponding to the eigenvalue 0.

Consider the homogeneously nonstationary ARMA$(2p, 2p)$ model,

$$\phi(B)y_t = \phi(B)v_t,$$

where $\phi(z) = -\phi_0 - \phi_1 z - \cdots - \phi_{2p}z^{2p}$, $\phi_0 = \phi_{2p} = -1$, $\phi_j = \phi_{2p-j}$, $j = 1, \ldots, p-1$. It is clear that all the roots of $\phi(z) = 0$ are on the unit circle. Assume that any root of $\phi(z) = 0$ is not real. This is called the *sinusoidal model*, which is important in spectral analysis. Let $\boldsymbol{\phi}^* = (\phi_0, \ldots, \phi_{2p})^t$. Then,

$$\Sigma(2p+1, 0)\boldsymbol{\phi}^* = \sigma^2\boldsymbol{\phi}^*.$$

Thus, σ^2 is an eigenvalue of $\Sigma(2p+1, 0)$ and $\boldsymbol{\phi}^*$ is the corresponding eigenvector. It is known that σ^2 is the smallest eigenvalue. The eigen-analysis is useful for estimating the order p and the parameters of the sinusoidal model. In this model we can calculate the eigenvalues using some recursive algorithms. Interested readers may consult the references in Section 5.9.

Because the trace is the sum of the eigenvalues, the trace of the covariance matrix may be used for ARMA model identification. (See, e.g., Young, Jakeman, and McMurtrie [1980] and Poskitt [1987].)

5.6.2 The SCAN Method

In multivariate time series analysis, the usefulness of canonical correlation analysis has been well espied. However, it is beyond the scope of this book to discuss it in detail. Interested readers may consult the references in Section 5.9.

We first consider the stationary case. Define a $(k+1) \times (k+1)$ matrix as

$$A(k,j) = \Sigma^{-1}(k+1,0)\Sigma(k+1,j+1)^t\Sigma^{-1}(k+1,0)\Sigma(k+1,j+1).$$

Let $\boldsymbol{w}_{k,t} = (y_t, \ldots, y_{t-k})^t$. Then, the number of zero canonical correlations of $\boldsymbol{w}_{k,t}$ and $\boldsymbol{w}_{k,t-j-1}$ are the number of zero eigenvalues of $A(k,j)$. Denote the smallest eigenvalue of $A(k,i)$ by $\mu(k,i)$. If $k = p, p+1, \ldots,$ $i = q, q+1, \ldots,$ then the smallest eigenvalue of $A(k,i)$ is 0 and $\boldsymbol{\phi}_*(k) = (\phi_0, \ldots, \phi_p, 0, \ldots, 0)^t$ is the corresponding eigenvector.

Property 5.26. *The SCAN Pattern*
Let $\{y_t\}$ be from the ARMA(p,q) model (1.1). Then the following holds.

1. $\mu(k,i) = 0, \quad (k,i) \in I_{p,q}$.
2. $\mu(p-1,i) \neq 0, \quad p \geq 1,\ i \geq \max(0, q-1)$.
3. $\mu(k,q-1) \neq 0, \quad q \geq 1,\ k \geq \max(0, p-1)$. □

Let $\hat{A}(k,i)$ be the estimate of $A(k,i)$ obtained by substituting the sample ACVF for the ACVF, and let $\hat{\mu}(k,i)$ be the smallest eigenvalue of $\hat{A}(k,i)$. Because the sample ACVF is consistent to the ACVF, $\hat{\mu}(k,i)$ is also consistent to $\mu(k,i)$.

For the nonstationary ARMA(p,q) model (1.3), we define

$$\hat{A}^*(k,i) = \hat{\boldsymbol{\beta}}^*(k,i+1)\hat{\boldsymbol{\beta}}(k,i+1).$$

Here $\hat{\boldsymbol{\beta}}(k,i+1)$ and $\hat{\boldsymbol{\beta}}^*(k,i+1)$ are $(k+1) \times (k+1)$ matrices defined by

$$\hat{\boldsymbol{\beta}}(k,i+1) = \left(\sum_{t=i+2+k}^{T} \boldsymbol{w}_{k,t-i-1}\boldsymbol{w}_{k,t-i-1}^t \right)^{-1} \left(\sum_{t=i+2+k}^{T} \boldsymbol{w}_{k,t-i-1}\boldsymbol{w}_{k,t}^t \right),$$

$$\hat{\boldsymbol{\beta}}^*(k,i+1) = \left(\sum_{t=i+2+k}^{T} \boldsymbol{w}_{k,t}\boldsymbol{w}_{k,t}^t \right)^{-1} \left(\sum_{t=i+2+k}^{T} \boldsymbol{w}_{k,t}\boldsymbol{w}_{k,t-i-1}^t \right).$$

Properties 1.1-1.4 imply that $\hat{\boldsymbol{\beta}}(k,i+1)$ and $\hat{\boldsymbol{\beta}}^*(k,i+1)$ exist for sufficiently large T. Thus, $\hat{A}^*(k,i)$ is defined for nonstationary ARMA processes. For stationary ARMA models, $\hat{A}(k,i)$ and $\hat{A}^*(k,i)$ are asymptotically equivalent. Let $\hat{\mu}^*(k,i)$ be the smallest eigenvalue of $\hat{A}^*(k,i)$. Then its limit has the same properties as $\mu(k,i)$.

Table 5.11. The SCAN array

$k \setminus i$	$\cdots$	$q-1$	q	$q+1$	$\cdots$
$\vdots$		$\vdots$	$\vdots$	$\vdots$	
$p-1$	$\cdots$	X	X	X	$\cdots$
p	$\cdots$	X	0	0	$\cdots$
$p+1$	$\cdots$	X	0	0	$\cdots$
$p+2$	$\cdots$	X	0	0	$\cdots$
$\vdots$		$\vdots$	$\vdots$	$\vdots$	

Note: X represents nonzero value.

Property 5.27. *The Generalized SCAN Pattern*
Let $\{y_1, \ldots, y_T\}$ be a T-realization from the ARMA(p,q) model (1.3). Let $\mu^*(k,i)$ be the limit of $\hat{\mu}^*(k,i)$ as $T \to \infty$. Then the following holds.

1. $\mu^*(k,i) = 0, \quad (k,i) \in I_{p,q}$.
2. $\mu^*(p-1,i) \neq 0, \quad p \geq 1,\ i \geq \max(0, q-1)$.
3. $\mu^*(k,q-1) \neq 0, \quad q \geq 1,\ k \geq \max(0, p-1)$. □

The two-way array of the SCAN is tabulated in Table 5.11. TT-85 defined a statistic as

$$c(k,i) = -(T-k-i)\ln\left\{1 - \frac{\hat{\mu}^*(k,i)}{\sum_{l=-i}^{i} r_l^2(z)}\right\}.$$

Here $r_l(z)$ is the sample autocorrelation at lag l of the process $\{z_{k,t}\}$, where

$$z_{k,t} = y_t - \hat{\phi}_1^{(i)} y_{t-1} - \cdots - \hat{\phi}_k^{(i)} y_{t-k}$$

and $(1, -\hat{\phi}_1^{(i)}, \ldots, -\hat{\phi}_k^{(i)})$ is a normalized eigenvector corresponding to $\hat{\mu}(k,i)$. Using the asymptotic distribution of canonical correlations (see, e.g., T. W. Anderson [1984, Section 13.5]), TT-85 presented the asymptotic distribution.

Theorem 5.15. Let $\{y_1, \ldots, y_T\}$ be a T-realization from the ARMA (p,q) model (1.3). If $(k,i) \in I_{p,q}$, then $c(k,i)$ is asymptotically χ_1^2 distributed. □

5.7 Woodside's Method

Woodside (1971) presented a pattern identification method for special ARMA models, which we discussed in Section 4.2. He considered the

ARMA(p,p) model

$$\phi(B)y_t = \theta(B)v_t,$$

where $\phi(B) = -\phi_0 - \phi_1 B - \cdots - \phi_p B^p$, $\theta(B) = -\theta_0 - \theta_1 B - \cdots - \theta_p B^p$ and $\phi_0 = \theta_0 = -1$. We assume that the ARMA model is stationary, invertible, and coprimal. The identification method is to use the covariance matrix $W(k)$ of

$$s(t,k) = (v_{t-1}, y_{t-1}, v_{t-2}, y_{t-2}, \ldots, v_{t-k}, y_{t-k})^t.$$

If we define the cross covariances of $\{y_t\}$ and $\{v_t\}$ by

$$\sigma_{yy}(k) = \text{cov}(y_t, y_{t+k}),$$

$$\sigma_{vv}(k) = \text{cov}(v_t, v_{t+k}),$$

$$\sigma_{vy}(k) = \text{cov}(v_t, y_{t+k}),$$

then the $2k \times 2k$ matrix $W(k)$ is

$$W(k) = \begin{pmatrix} \sigma_{vv}(0) & \sigma_{vy}(0) & \cdots & \sigma_{vv}(-k+1) & \sigma_{vy}(-k+1) \\ \sigma_{yv}(0) & \sigma_{yy}(0) & \cdots & \sigma_{yv}(-k+1) & \sigma_{yy}(-k+1) \\ \vdots & \vdots & & \vdots & \vdots \\ \sigma_{vv}(k-1) & \sigma_{vy}(k-1) & \cdots & \sigma_{vv}(0) & \sigma_{vy}(0) \\ \sigma_{yv}(k-1) & \sigma_{yy}(k-1) & \cdots & \sigma_{yv}(0) & \sigma_{yy}(0) \end{pmatrix}.$$

Clearly, $\{W(k) \mid k = 1, 2, \ldots\}$ is a sequence of nested matrices. The EYW equations imply the following property.

Property 5.28. *The W Pattern.*
Let $\{y_t\}$ be from the ARMA(p,p) model. If $k \geq p$, then the rank of $W(k)$ is $k+p$. □

The idea of using the matrix W is originally due to Lee (1964, pp. 98-111). Woodside utilized this property for model identification by comparing the determinants of $W(k)$, investigating the zeroness of the smallest eigenvalues of $W(k)$, comparing residuals and computing the likelihood ratio test statistics. To calculate the determinants and the inverses of $\{W(k) \mid k = 1, 2, \ldots\}$, we can apply Algorithm 4.2. In control theory and signal processing there are some progresses on Woodside's method. Interested readers may consult the references in Section 5.9.

5.8 Three Systems of Equations

The previous pattern identification methods are mostly based on the EYW equations. There are three other possible systems of equations for ARMA model identification.

The first possibility is to use the IACF of the ARMA(p, q) process. The invertibility assumption implies the IEYW equations, which are defined in Section 2.2:

$$\theta_1 \rho i(p-1) + \cdots + \theta_q \rho i(p-q) \neq \rho i(p),$$
$$\theta_1 \rho i(j-1) + \cdots + \theta_q \rho i(j-q) = \rho i(j), \quad j = p+1, p+2, \ldots .$$

Then,

$$Bi(q,p)\boldsymbol{\theta} = \boldsymbol{\rho i}(q,p),$$

where $Bi(k,i)$ and $\boldsymbol{\rho i}(k,i)$ are defined by

$$Bi(k,i) = \begin{pmatrix} \rho i(i) & \rho i(i-1) & \cdots & \rho i(i-k+1) \\ \rho i(i+1) & \rho i(i) & \cdots & \rho i(i-k+2) \\ \vdots & \vdots & & \vdots \\ \rho i_(i+k-1) & \rho i(i+k-2) & \cdots & \rho i(i) \end{pmatrix},$$

$$\boldsymbol{\rho i}(k,i) = (\rho i(i+1), \ldots, \rho i(i+k))^t.$$

Let $Di(k,i)$ be the determinant of $Bi(k,i)$, and define an index set as

$$I_{q,p} = \{(k,p) \mid k = q, q+1, \ldots\} \bigcup \{(q,i) \mid i = p, p+1, \ldots\}.$$

Then the IEYW equations imply the following property.

Property 5.29. *The Di Pattern*
A stochastic process has the ARMA(p, q) representation (1.1) if and only if q and p are the smallest integers satisfying

$$Di(k,i) = 0, \quad (k,i) \in I_{q,p}. \quad \square$$

The second possible system of equations uses the MA(∞) representation of the ARMA(p, q) process. The stationarity assumption implies the ARMA process can be represented by

$$y_t = \sum_{j=0}^{\infty} \psi_j v_{t-j},$$

where

$$\psi(z) = \phi^{-1}(z)\theta(z) = \sum_{j=0}^{\infty} \psi_j z^j.$$

It implies

$$\phi_1 \psi_{q-1} + \cdots + \phi_p \psi_{q-p} \neq \psi_q,$$
$$\phi_1 \psi_{j-1} + \cdots + \phi_p \psi_{j-p} = \psi_j, \quad j = q+1, q+2, \ldots,$$

where $\psi_j = 0,\ j < 0$. Thus,

$$\Psi(p,q)\boldsymbol{\phi}(p,q) = \boldsymbol{\psi}(p,q),$$

where $\Psi(k,i)$ and $\boldsymbol{\psi}(k,i)$ are defined by

$$\Psi(k,i) = \begin{pmatrix} \psi_i & \psi_{i-1} & \cdots & \psi_{i-k+1} \\ \psi_{i+1} & \psi_i & \cdots & \psi_{i-k+2} \\ \vdots & \vdots & & \vdots \\ \psi_{i+k-1} & \psi_{i+k-2} & \cdots & \psi_i \end{pmatrix},$$

$$\boldsymbol{\psi}(k,i) = (\psi_{i+1}, \ldots, \psi_{i+k})^t.$$

Property 5.30. *The* det(Ψ) *Pattern*
A stochastic process has the ARMA(p,q) representation (1.1) *if and only if* p and q are the smallest integers satisfying

$$\det(\Psi(k,i)) = 0, \quad (k,i) \in I_{p,q}. \quad \square$$

It can easily be shown that

$$\text{cov}(y_{t+j}, v_t) = \sigma^2\psi_j, \quad j = 0, 1, \ldots.$$

Thus, properties of the Ψ array can be studied through the cross-correlation of the observations $\{y_t\}$ and the estimated innovations $\{\hat{v}_t\}$,

$$r_{vy}(j) = \frac{\sum_{t=1}^{T-j} \hat{v}_t y_{t+j}}{\left\{\sum_{t=1}^{T} \hat{v}_t^2 \sum_{t=1}^{T-j} y_{t+j}^2\right\}^{1/2}}, \quad j = 0, 1, \ldots.$$

Hokstad (1983) has proposed to use the estimated cross-correlation function for diagnostic checking of the selected model, which can be regarded as an estimate of the normalized λ function.

The third possible system of equations utilizes the AR(∞) representation of the ARMA(p,q) process. The invertibility assumption implies that the ARMA process can be represented as

$$y_t = \sum_{j=1}^{\infty} \pi_j y_{t-j} + v_t,$$

where

$$\pi(z) = \phi^{-1}(z)\theta(z) = -\sum_{j=0}^{\infty} \pi_j z^j, \quad \pi_0 = -1.$$

Then,

$$\theta_1\pi_{p-1} + \cdots + \theta_q\pi_{p-q} \neq \pi_p,$$
$$\theta_1\pi_{j-1} + \cdots + \theta_q\pi_{j-q} = \pi_j, \quad j = p+1, p+2, \ldots,$$

where $\pi_j = 0,\ j < 0$. Therefore,

$$\begin{pmatrix} \pi_p & \pi_{p-1} & \cdots & \pi_{p-q+1} \\ \pi_{p+1} & \pi_p & \cdots & \pi_{p-q+2} \\ \vdots & \vdots & & \vdots \\ \pi_{p+q-1} & \pi_{p+q-2} & \cdots & \pi_p \end{pmatrix} \begin{pmatrix} \theta_1 \\ \theta_2 \\ \vdots \\ \theta_q \end{pmatrix} = \begin{pmatrix} \pi_{p+1} \\ \pi_{p+2} \\ \vdots \\ \pi_{p+q} \end{pmatrix}.$$

Let $\Pi(k,i)$ be the i-dimensional Toeplitz matrix, whose (r,s) element is π_{k+r-s}. Then $\det(\Pi(k,i))$ has a useful pattern for ARMA model identification as follows.

Property 5.31. *The* $\det(\Pi)$ *Pattern*
A stochastic process has the ARMA(p,q) representation (1.1) *if and only if* p and q are the smallest integers satisfying

$$\det(\Pi(k,i)) = 0, \quad (k,i) \in I_{q,p}.. \quad \square$$

Graupe, Krause, and Moore (1975) have already utilized the $\det(\Psi)$ pattern. However, any statistical inference tool about it has not yet been provided.

Even though we mentioned only the Corner-array type pattern identification methods of the three systems of equations, it is possible to apply any of the pattern identification methods to each of the three systems. The present author considers the patterns from the second and the third systems to be better than the ones based on the EYW equations, for they depend not only on the AR parameters but also on the MA parameters.

5.9 Additional References

- (Section 5.1) A detailed proof of Property 5.1 can be found in Beguin, Gourieroux and Monfort (1980). In the case of vector ARMA processes, the correlation matrix becomes a block Toeplitz matrix as discussed in Section 5.2. An, Chen, and Hannan (1983) studied the rank of the block matrix.

- (Section 5.2) The R and S array method is based on Gray, Houston and Morgan's suggestion (1978). For the Pye-Atchison algorithm, refer to Bednar and Roberts (1985). The R and S array method has been used to detect the nonstationarity factor on the unit circle and to examine seasonal data by Gray and Woodward (1981) and Woodward and Gray (1981). Tucker (1982) has shown that the R and S array method has much wider applications than the ARMA model identification such as distinguishing between autoregressive integrated moving-average (ARIMA) and signal plus ARMA noise processes.

- (Section 5.3) About the relation between the D array and the Padé approximation, readers may refer to Baker (1975), Tucker (1982), and Petrushev and Popov (1987, pp. 329-347). Also, refer to J. C. Chow (1972), Lindberger (1973), and Solo (1986a, 1986b) for more details of the D pattern.
- (Section 5.4) Jenkins and Alavi (1981) named $\phi_{k,k}^{(i)}$ the kth order i-conditioned partial correlation and used the GPAC for identifying vector ARMA models. A similar proposal was mentioned by Tiao and Box (1981) for multivariate applications. Gooijer and Saikkonen (1988) proposed the use of the GPAC with an efficient estimation method due to Pham (1979).
- (Section 5.5) For more details about the ESACF method, readers may refer to Tiao and Tsay (1983b, 1989), Tsay (1989), and Tsay and Tiao (1990). Pham (1984a, 1988) has presented an estimation method of AR parameters similar to the ESACF and an identification method alike to the ESACF procedure.
- (Section 5.6) For the sinusoidal model, refer to Pisarenko (1973), Ulrych and Clayton (1976), Quinn (1986, 1989), and the references therein.
- (Section 5.6) We can calculate the eigenvalues of symmetric Toeplitz matrices using recursive algorithms by Cybenko and Van Loan (1986), Wilkes and Hayes (1987) and Trench (1989). Stoica and Nehorai (1988) applied the Levinson-Durbin algorithm to the sinusoidal model.
- (Section 5.6) About the canonical correlation analysis of multivariate time series, readers may refer to Brillinger (1981, pp. 367-391), Akaike (1975, 1976), Box and Tiao (1977), Hannan (1979), Cooper and Wood (1982), Tjøstheim and Paulsen (1982), Velu, Reinsel, and Wichern (1986), Hannan and Poskitt (1988), Velu, Wichern, and Reinsel (1987), Tsay (1989), Tiao and Tsay (1989), Swift (1990), Vaninskii and Yaglom (1990), Aoki (1990), and the references therein.
- (Section 5.7) In control theory and signal processing there are some progresses of Woodside's method. However, the models adopted are far more complicated than the ARMA model. Interested readers may refer to Wellstead (1978), Young, Jakeman, and McMurtrie (1980), Young (1985), and the references therein.

6

Testing Hypothesis Methods

Historically speaking, the hypothesis testing methods had been dominantly used to choose tentative ARMA models until Box-Jenkins' identification method appeared. Nowadays they are primarily used for testing model inadequacy after choosing the orders and estimating the parameters, which Box and Jenkins (1976) called the *model diagnostic checking.* Therefore, we discuss them in this final chapter.

6.1 Three Asymptotic Test Procedures

Hypothesis testing concerns the question of whether the assumed model is fitted to the observations or not. Let $L(\boldsymbol{\tau}, \boldsymbol{y})$ be the likelihood function of $\boldsymbol{y} = (y_1, \ldots, y_T)^t$, where $\boldsymbol{\tau}$ is the parameter vector belonging to the parameter space $\Omega \subset R^K$. We assume that the likelihood function satisfies the regularity conditions necessary to prove the asymptotic normality of the ML estimates. We define a subspace ω as

$$\omega = \{\boldsymbol{\tau} \mid \boldsymbol{r}(\boldsymbol{\tau}) = \boldsymbol{0}\}.$$

Here $\boldsymbol{r}(\boldsymbol{\tau})$ is a k-dimensional differentiable function ($k < K$) and has no redundant constraints. We consider testing the null hypothesis

$$H_o: \ \boldsymbol{\tau} \in \omega.$$

Denote the log-likelihood function by $l(\boldsymbol{\tau}) = \ln L(\boldsymbol{\tau}, \boldsymbol{y})$, the unconstrained ML estimate of $\boldsymbol{\tau}$ in Ω by $\check{\boldsymbol{\tau}}$, and the constrained ML estimate in ω by $\tilde{\boldsymbol{\tau}}$. Also, denote the score and the Fisher information matrix divided by T as

$$\boldsymbol{s}(\boldsymbol{\tau}) = \frac{\partial l(\boldsymbol{\tau})}{\partial \boldsymbol{\tau}},$$

$$\mathcal{J}(\boldsymbol{\tau}) = -\frac{1}{T} E\left(\frac{\partial^2 l(\boldsymbol{\tau})}{\partial \boldsymbol{\tau} \partial \boldsymbol{\tau}^t}\right),$$

respectively.

We will discuss three asymptotic tests of H_o, i.e., the Wald test, the likelihood ratio (LR) test, and the Lagrange multiplier (LM) test. For more details on these three tests, readers may consult the references in Section 6.5.

The Wald (1943) test statistic is defined by

$$W = T\boldsymbol{r}(\check{\boldsymbol{\tau}})^t \left\{ \frac{\partial \boldsymbol{r}(\check{\boldsymbol{\tau}})}{\partial \boldsymbol{\tau}^t} \mathcal{J}^{-1}(\check{\boldsymbol{\tau}}) \frac{\partial \boldsymbol{r}(\check{\boldsymbol{\tau}})^t}{\partial \boldsymbol{\tau}} \right\}^{-1} \boldsymbol{r}(\check{\boldsymbol{\tau}}).$$

If $\mathcal{J}(\check{\boldsymbol{\tau}})$ is consistent to $\mathcal{J}(\boldsymbol{\tau})$, then W is asymptotically χ_k^2 distributed under the null hypothesis.

The LR test statistic is defined by

$$LR = 2\left\{l(\tilde{\boldsymbol{\tau}}) - l(\check{\boldsymbol{\tau}})\right\},$$

which is also referred to as Wilks' test statistic (1932, 1938). It is known to be asymptotically optimal for many problems. (See, e.g., Lehmann [1959, Section 7.13] and T. W. Anderson [1984, Chapter 8].) The LR test is asymptotically χ_k^2 distributed under the null hypothesis. Durbin (1970) has presented a theorem applicable to the derivation of the LR test statistic when the ML estimation in ω is much easier than in Ω.

The LM test is to maximize the log-likelihood subject to the null hypothesis $\boldsymbol{r}(\boldsymbol{\tau}) = \boldsymbol{0}$, which was proposed by Aitchison and Silvey (1958, 1960) and Silvey (1959). Rao (1947) proposed the same test and called it the score test. Consider the Lagrangean function

$$l(\boldsymbol{\tau}) + \boldsymbol{\lambda}^t \boldsymbol{r}(\boldsymbol{\tau}).$$

The first-order conditions are

$$\begin{aligned} \boldsymbol{s}(\boldsymbol{\tau}) + \boldsymbol{\lambda}^t \frac{\partial \boldsymbol{r}(\boldsymbol{\tau})}{\partial \boldsymbol{\tau}} &= \boldsymbol{0}, \\ \boldsymbol{r}(\boldsymbol{\tau}) &= \boldsymbol{0}. \end{aligned}$$

If the null hypothesis is correct, then the restricted estimate has a tendency to be near the unrestricted estimate, and then the score $\boldsymbol{s}(\tilde{\boldsymbol{\tau}})$ is close to the zero vector. In other words, if the Lagrange multiplier, which is also called the shadow price, is large in some metric senses, then the constraint, i.e., the null hypothesis, should be rejected. Based on the first-order conditions, we may estimate the squared norm of the Lagrange multiplier by

$$\mathrm{LM} = \frac{1}{T} \boldsymbol{s}(\tilde{\boldsymbol{\tau}})^t \mathcal{J}^{-1}(\tilde{\boldsymbol{\tau}}) \boldsymbol{s}(\tilde{\boldsymbol{\tau}}),$$

which is the definition of the LM test statistic. It is also asymptotically χ_k^2 distributed under the null hypothesis.

The three statistics are discrimination measures between the null and the alternative hypotheses. They can be geometrically interpreted. If $k = 1$, then the Wald statistic and the LR statistic represent the horizontal and the vertical distances in the XY coordinate plane, respectively. The LM statistic represents the difference between the slopes. We can generalize the geometrical interpretation to $k > 1$. If $l(\boldsymbol{\tau})$ is a quadratic function, they yield exactly the same test. In addition, if $l(\boldsymbol{\tau})$ is well approximated by a quadratic function, they will be similar. Under mild regularity conditions, the three test statistics have the same asymptotic distributions not only under the null hypothesis but also under the alternative hypothesis

$$H_a : \boldsymbol{r}(\boldsymbol{\tau}) = \boldsymbol{O}\left(\frac{1}{\sqrt{T}}\right).$$

(See, e.g., Engle [1984].) However, the three statistics are different for small sample cases. For the linear model and the multivariate linear regression, Savin (1976), Berndt and Savin (1977), and Breusch (1979) have shown that

$$\mathrm{W} \geq \mathrm{LR} \geq \mathrm{LM}.$$

These inequalities imply that if the LM test rejects the null hypothesis, so do the Wald and the LR tests. However, it should be noted that the inequalities do not say anything about the relative values of the tests. What to choose among the three tests depends on criteria such as small sample properties and computational convenience. In the case where the alternative model is very general, i.e., it contains much more parameters than the null model, we prefer the LM method, for it does not require estimation of the alternative model. In this case it is expected that the LM test depends little on the precise form of the alternative model. In other words, when the alternative model is broad, we can expect the LM test to resemble a pure significance test, which is a test based on a statistic whose behavior is considered only under the null hypothesis without taking into account the alternative (Cox and Hinkley [1974, pp. 64-82]). We might find a general alternative model against which the LM test is equivalent to the given pure significance test (Hosking [1980b]). Basawa, Huggins, and Staudte (1985) presented robust analogues of the Wald and the LM statistics for testing composite hypotheses appearing in time series analysis.

6.2 Some Test Statistics

We first consider the case where a T-realization $\{y_1, \ldots, y_T\}$ is from an AR model and the null hypothesis is

$$H_o : \phi_{k+1} = \cdots = \phi_{k+m} = 0.$$

A test statistic based on the PACF is

$$C_1 = T \sum_{j=k+1}^{k+m} \hat{\phi}_{j,j}^2.$$

Theorem 2.1 implies that it is asymptotically χ_m^2 distributed under the null hypothesis.

Quenouille (1947) proposed a statistic

$$C_2 = \frac{1}{\tilde{\sigma}^4} \sum_{j=k+1}^{k+m} \tilde{h}_j^2,$$

where

$$\tilde{h}_j = \frac{1}{\sqrt{T}} \sum_{l=0}^{k} \sum_{n=0}^{k} \tilde{\phi}_{k,l} \tilde{\phi}_{k,n} \sum_{t=1}^{T} y_{t-j+l} y_{t-n}.$$

We may define a test statistic by substituting the YW estimates for the ML estimates in C_2 as

$$C_3 = \frac{1}{\hat{\sigma}^4} \sum_{j=k+1}^{k+m} \hat{h}_j^2,$$

where

$$\hat{h}_j = \frac{1}{\sqrt{T}} \sum_{l=0}^{k} \sum_{n=0}^{k} \hat{\phi}_{k,l} \hat{\phi}_{k,n} \sum_{t=1}^{T} y_{t-j+l} y_{t-n}.$$

Under the null hypothesis, C_2 and C_3 are asymptotically χ_m^2 distributed. Bartlett and Rajalakshman (1953) have presented a simple derivation of Quenouille's test. Walker (1952) has presented the asymptotic power of Quenouille's test and has shown that it is asymptotically equivalent to the LR test. Hosking (1978) has shown that Quenouille's test equals the LM test for the hypotheses

$$H_o: \; \phi_k(B) y_t = v_t$$

versus

$$H_a: \; \phi_k(B) y_t + \sum_{j=1}^{m} \lambda_j \phi_k(B^{-1}) y_{t-k-j} = v_t.$$

The Wald statistic for testing H_o can be found in T. W. Anderson (1971, p. 216). Partition the correlation matrix estimate $\hat{B}(k+m,0)$ into k and m rows and columns and the parameter vector estimate $\hat{\phi}(k+m,0) = (\hat{\phi}_{k+m,1}, \ldots, \hat{\phi}_{k+m,k+m})^t$ correspondingly as

$$\hat{B}(k+m,0) = \begin{pmatrix} \hat{B}(k,0) & \hat{B}_{12} \\ \hat{B}_{21} & \hat{B}(m,0) \end{pmatrix}, \quad \hat{\phi}(k+m,0) = \begin{pmatrix} \hat{\phi}_1 \\ \hat{\phi}_2 \end{pmatrix}.$$

Then the Wald statistic is

$$C_4 = T \frac{1}{\hat{\sigma}_{k+m}^2} \hat{\phi}_2^t \left\{ \hat{B}(m,0) - \hat{B}_{21} \hat{B}^{-1}(k,0) \hat{B}_{12} \right\} \hat{\phi}_2,$$

which is asymptotically χ_m^2 distributed under the null hypothesis. It is known (see, e.g., T. W. Anderson [1971, pp. 214-223]) that as $T \to \infty$, the statistics C_1, C_2, C_3, and C_4 are asymptotically equivalent in probability under the null hypothesis.

Whittle (1952a, 1954) presented the LR statistic

$$C_5 = (T - k - m) \ln \frac{\tilde{\sigma}_k^2}{\tilde{\sigma}_{k+m}^2}.$$

A Taylor expansion implies that it is approximately

$$C_6 = (T - k - m) \frac{\tilde{\sigma}_k^2 - \tilde{\sigma}_{k+m}^2}{\tilde{\sigma}_{k+m}^2}.$$

Another form of the LR test is

$$C_7 = (T - k - m) \frac{\hat{R}^2_{k+m+1\cdot 1,\ldots,k+m} - \hat{R}^2_{k+1\cdot 1,\ldots,k}}{1 - \hat{R}^2_{k+m+1\cdot 1,\ldots,k+m}},$$

where $\hat{R}^2_{r+1\cdot 1,\ldots,r}$ is the sample multiple correlation coefficient of y_{t+r+1} and $y_{t+r}, \ldots, y_{t+1}$. Under the null hypothesis the LR test is asymptotically equivalent to

$$-(T - k - m) \sum_{j=k+1}^{k+m} \ln(1 - \tilde{\phi}_{j,j}^2) \approx (T - k - m) \sum_{j=k+1}^{k+m} \tilde{\phi}_{j,j}^2,$$

for sufficiently large T. Thus, C_5, C_6 and C_7 are asymptotically χ_m^2 distributed under the null hypothesis. McClave (1975) has applied the statistic C_6 to a stepwise search of the AR order, which is akin to the stepwise regression and is called the max χ^2 method.

Godfrey (1978a) has derived the LM tests for randomness, i.e., the white noise model against the AR model and against the MA model. He has shown that the test against the AR(p) model is exactly the same as that against the MA(p) model. Tanaka (1983) presented the LM test of the AR(p) model against the AR(p) model with measurement error.

Wold (1949) and Walker (1950) extended Quenouille's test to an MA model and a mixed ARMA model, respectively. Walker's statistic to test the null hypothesis that $y_1, \ldots, y_T$ are from an ARMA(k, i) model is

$$C_8 = T\tilde{\boldsymbol{h}}^t(k,i) W_m^{-1}(k,i) \tilde{\boldsymbol{h}}(k,i),$$

where $W_m(k,i)$ is the same covariance matrix as that of the E and J statistics defined in Section 5.1 and

$$\tilde{\boldsymbol{h}}(k,i) = (\tilde{h}_1(k,i), \ldots, \tilde{h}_m(k,i))^t,$$

$$\tilde{h}_j(k,i) = \frac{1}{\hat{\sigma}(0)} \sum_{l=0}^{k} \sum_{n=0}^{k} \tilde{\phi}_{k,l}^{(i)} \tilde{\phi}_{k,n}^{(i)} \hat{\sigma}(j + k + i - l - n).$$

It is asymptotically χ^2_m distributed under the null hypothesis. Bartlett and Diananda (1950) presented a simple but general method of its derivation using a triangular transformation due to Wold (1949). Choi (1991g) has shown that Walker's statistic is asymptotically equivalent to the E and the J statistics. Hosking (1980b) has shown that Walker's test statistic is the LM statistic to test

$$H_o: \ \phi_k(B)y_t = \theta_i(B)v_t$$

versus

$$H_a: \ \phi_k(B)y_t + \sum_{j=1}^{m} \lambda_i \phi_k(B^{-1})\theta_i(B^{-1})\theta_i(B)y_{t-k-i-j} = \theta_i(B)v_t,$$

where $\phi_k(z)$ and $\theta_i(z)$ are the kth and the ith polynomials, respectively.

Against the null hypothesis

$$H_o: \ \phi_k(B)y_t = \theta_i(B)v_t,$$

we may consider three possible alternative hypotheses:

$$H_A: \ \phi_k(B)\left\{y_t + \sum_{j=1}^{m} \lambda_i y_{t-j}\right\} = \theta_i(B)v_t,$$

$$H_B: \ \phi_k(B)y_t = \theta_i(B)\left\{v_t + \sum_{j=1}^{n} \lambda_i v_{t-j}\right\},$$

$$H_C: \ \phi_k(B)\left\{y_t + \sum_{j=1}^{m} \lambda_i y_{t-j}\right\} = \theta_i(B)\left\{v_t + \sum_{j=1}^{n} \lambda_i v_{t-j}\right\}.$$

Godfrey (1978b, 1979), Hosking (1981a), Pham (1986, 1987) and Ljung (1988) derived the LM tests for the hypotheses H_o versus H_A and for H_o *versus* H_B. Poskitt and Tremayne (1981a, 1981b, 1986) and Pötscher (1985) generalized the LM method to vector ARMA models and to transfer function models, respectively.

In some circumstances time series analysts wish to test the null hypothesis H_o against the alternative hypothesis H_C. As discussed in Chapter 4, the alternative leads to a singularity problem. Fitts (1973) has considered the problem of fitting observations from a white noise process to the ARMA$(1,1)$ model. The likelihood function does not have the unique maximum with respect to ϕ_1 and θ_1, but it is asymptotically maximized anywhere along the line $\phi_1 = \theta_1$. Fitts called it perfectly correlated. This property implies the singularity of the covariance matrix of the estimates. Hannan (1970a, pp. 388-389, 409-414) has shown that the covariance matrix of the ML estimates of an ARMA(k,i) model is singular if $\phi_k = \theta_i = 0$. There are more references about the singularity problem in Section 6.5.

Taniguchi (1985, 1988) presented asymptotic expansions of the distributions of the W, the LM, and the LR test statistics for Gaussian ARMA processes. We can make some ARMA testing procedures based on Kolmogorov's identity

$$\sigma^2 = (2\pi)\exp\left\{\frac{1}{2\pi}\int_{-\pi}^{\pi} \ln S(\lambda)d\lambda\right\},$$

where $S(\lambda)$ is the spectral density of the ARMA process. Because σ^2 is the variance of the best linear prediction error (see, e.g., Hannan [1970a, pp. 137-140]), it is natural to choose a model minimizing its estimate. Interested readers may consult the references in Section 6.5.

6.3 The Portmanteau Statistic

After the orders are selected and the parameters are estimated, we usually check whether the chosen model is adequate or not. A standard approach to model diagnostic checking is based on the residual ACRF. If $v_1, \ldots, v_T$ are independent and identically distributed, then Theorem 1.2 implies that

$$\boldsymbol{\rho}(v) = \left(\frac{\sum_{t=1}^{T-1} v_t v_{t+1}}{\sum_{t=1}^{T} v_t^2}, \ldots, \frac{\sum_{t=1}^{T-m} v_t v_{t+m}}{\sum_{t=1}^{T} v_t^2}\right)^t$$

is asymptotically normally distributed with mean $\mathbf{o}$ and covariance matrix $(1/T)I$. Let $\hat{v}_1, \ldots, \hat{v}_T$ be the residuals due to the selected ARMA(k, i) model. The residual autocorrelation at lag j is defined by

$$r_j(v) = \frac{\sum \hat{v}_t \hat{v}_{t+j}}{\sum \hat{v}_t^2}, \quad j = 1, 2, \ldots.$$

Box and Pierce (1970) and McLeod (1978) have shown that if the selected model is correct, then

$$\boldsymbol{r}(v) = (r_1(v), \ldots, r_m(v))^t$$

has asymptotically the same distribution as $(I - P)\boldsymbol{\rho}(v)$, where P is an idempotent matrix with rank$(P) = k + i$. Box and Pierce proposed a test statistic using $\boldsymbol{r}(v)$ to test model adequacy,

$$Q_m = T\sum_{j=1}^{m} r_j^2(v),$$

which is called the portmanteau statistic. Cochran's theorem implies that Q_m is asymptotically χ^2_{m-k-i} distributed if the model is correct. Davies, Triggs, and Newbold (1977) have reported that for moderate sample sizes the mean and the variance of Q_m can differ substantially from those by the asymptotic theory. Particularly, the mean is far too low. In such cases the true significance levels are often very much lower than those given by the χ^2 approximation. There are more references about the performance of the Q_m statistic in Section 6.5.

We can modify the Q_m statistic using the refined variance of the sample ACRF

$$\operatorname{var}\left(\frac{\sum_{t=1}^{T-j} v_t v_{t+j}}{\sum_{t=1}^{T} v_t^2}\right) = \frac{T-j}{T(T+2)}, \quad j = 1, 2, \ldots,$$

by R. L. Anderson (1942) and Moran (1948). (Also, refer to Merikoski and Pukkila [1983].) The modified portmanteau statistic is

$$Q_m^* = T(T+2)\sum_{j=1}^{m} \frac{1}{T-j} r_j^2(v),$$

which is also called the Box-Pierce-Ljung statistic. Hosking (1978, 1980a, 1980b, 1981a) has shown that the portmanteau statistic equals the LM test statistic (or equivalently the LR test statistic) for testing H_o against either H_A or H_B, which are defined in the previous subsection. There are more references about it in Section 6.5.

Ljung and Box (1978) have suggested using Q_m^* instead of Q_m, for the distribution of the Q_m^* statistic is closer to the χ^2_{m-k-i} distribution than that of the Q_m statistic. Godfrey (1979) has argued that the finite sample significance levels of the Q_m^* statistic are likely to be closer to the asymptotically achieved values than those of the Q_m statistic. However, Davies, Triggs, and Newbold (1977) have shown that the mean of Q_m^* is closer to $m - k - i$ than that of Q_m, but its variance is larger than that of Q_m. They have concluded that time series analysts would rather use other diagnostic checking methods such as overfitting or examination of the first two or three autocorrelation terms of the residuals. There are more references about the Q_m^* statistic in Section 6.5.

To test the model acceptability using the Q_m^* statistic, it is necessary to specify m. Usually it has been selected as high as 20 or 30 even when a low order model might be more appropriate. Ljung (1986) has shown that the properties of the Q_m^* statistic can be improved by choosing a smaller value of m. The multivariate portmanteau statistic has been discussed by Chitturi (1974, 1976), Hosking (1980c, 1981a, 1981b), Li and McLeod (1981) and Li (1985). Petruccelli and Davies (1986) presented a portmanteau test for self-exciting threshold autoregressive-type nonlinearity. McLeod and Li (1983) have asserted that there are some hydrological and economic time series in which the squared residuals of the best fitting ARMA model are significantly autocorrelated even though the usual residual autocorrelations do not suggest any model inadequacy. Then, they have presented the asymptotic joint distribution of the normalized squared residual autocorrelations and the corresponding portmanteau statistic.

To test the randomness of the residuals, Knoke (1975, 1977) presented an LM test and O'Brien (1980) proposed an LR test. Also, Kedem and Slud (1981) proposed a nonparametric test statistic.

6.4 Sequential Testing Procedures

The hypothesis testing problem in the previous section is either that the AR model has order less than k given the condition its degree is at most $k+m$ or that the ARMA model has orders less than k and i given the condition they are at most either $k+m$ and i or k and $i+m$. In order to determine the orders of an ARMA process using the test statistics in the previous section, we confront a multiple-decision problem.

We first consider a pure AR model. Assume that the true AR order p is unknown and a maximum order K is specified. Then a method to determine the order is to test the following null hypotheses sequentially. (See, e.g., T. W. Anderson [1963b; 1971, pp. 34-46, 260-276].)

$$\begin{array}{lcl} H_K & : & \phi_K = 0, \\ H_{K-1} & : & \phi_K = \phi_{K-1} = 0, \\ \vdots & & \vdots \\ H_1 & : & \phi_K = \phi_{K-1} = \cdots = \phi_1 = 0. \end{array}$$

If any of these hypotheses is rejected, then all of the following hypotheses are rejected. We can use the test statistics in the previous section to test each null hypothesis. However, it should be noted that Type I error in this sequential testing procedure differs from the significance level chosen for the individual test.

Jenkins and Alavi (1981) and Tiao and Box (1981) proposed another sequential testing procedure.

$$\begin{array}{lcl} H_K & : & \phi_K = 0, \\ H_{K-1} & : & \phi_{K-1} = 0, \\ \vdots & & \vdots \\ H_1 & : & \phi_1 = 0. \end{array}$$

We can test each null hypothesis H_k using a test statistic $T\hat{\phi}_{k,k}^2$, which is equivalent to Box-Jenkins' identification test statistic.

We may choose the order of an MA model by replacing the AR parameters with the MA parameters and by applying the sequential procedures. However, in the case of a mixed ARMA process it is difficult to decide the priorities among the null hypotheses. One of the possible ways is to determine the order $m = \max(k, i)$ first, and then to choose k and i among $\{(k,i) \mid k = m, i = 0, \ldots, m\} \bigcup \{(k,i) \mid k = 0, \ldots, m, i = m\}$ as the HR method. Another possibility is to approximate the ARMA process by either an AR model or an MA model. Pötscher (1983) has proposed a stepwise testing procedure using LM tests for ARMA model identification and has shown that the selected orders are strongly consistent.

The sequential procedures have difficulty in controlling overall Type I error and they are computationally very expensive as tentative model identification. Moreover, the sequential test procedures may result in selecting too high orders as the backward regression procedure does.

6.5 Additional References

- (Section 6.1) For more details about the Wald, the LR, and the LM tests, readers may refer to Breusch and Pagan (1980), Berndt and Savin (1977), Engle (1984), and the references therein.
- (Section 6.2) About the singularity problem of overparameterization of the ARMA model, refer to Godfrey (1978a), Breusch (1978), Shibata (1986), and Veres (1987). Poskitt and Tremayne (1980) presented a ridge-regression type LM test of H_o against H_C.
- (Section 6.2) For ARMA modeling methods utilizing Kolmogorov's identity, refer to Cameron (1978), Milhøj (1981), Saikkonen (1983), Pham (1986, 1987), and the references therein.
- (Section 6.3) About the performance of the Q_m and the Q_m^* statistics, refer to Prothero and Wallis (1976), Davies and Newbold (1979), Ansley and Newbold (1979), Godolphin (1980b), and Clarke and Godolphin (1982).
- (Section 6.3) The equivalence of the portmanteau statistic and the LM statistic is also shown by Newbold (1980) and Poskitt and Tremayne (1981b, 1982).

Bibliography

[1] Ables, J. G. (1974). Maximum entropy spectral analysis, *Astronomy Astrophysics: Supplement Ser.*, **15**, 383-393.

[2] Abraham, B. and J. Ledolter (1984). A note on inverse autocorrelations, *Biometrika*, **71**, 609-614.

[3] Abraham, B. and J. Ledolter (1986). Forecast functions implied by autoregressive integrated moving average models and other related forecast procedures, *Inter. Statistical Review*, **54**, 51-66.

[4] Ahtola, J. and G. C. Tiao (1984). Parameter inference for a nearly nonstationary first-order autoregressive model, *Biometrika*, **71**, 263-272.

[5] Ahtola, J. and G. C. Tiao (1987a). Distributions of least squares estimators of autoregressive parameters for a process with complex roots on the unit circle, *J. Time Series Anal.*, **8**, 1-14.

[6] Ahtola, J. and G. C. Tiao (1987b). A note on asymptotic inference in autoregressive models with roots on the unit circle, *J. Time Series Anal.*, **8**, 15-19.

[7] Aitchison, J. and S. D. Silvey (1958). Maximum-likelihood estimation of parameters subject to restraints, *Ann. Math. Statist.*, **29**, 813-828.

[8] Aitchison, J. and S. D. Silvey (1960). Maximum-likelihood estimation procedures and associated tests of significance, *J. Roy. Statist. Soc. Ser. B*, **22**, 154-171.

[9] Akaike, H. (1969a). Fitting autoregressive models for prediction, *Ann. Inst. Statist. Math.*, **21**, 243-247.

[10] Akaike, H. (1969b). Power spectrum estimation through autoregressive model fitting, *Ann. Inst. Statist. Math.*, **21**, 407-419.

[11] Akaike, H. (1970a). Statistical predictor identification, *Ann. Inst. Statist. Math.*, **22**, 203-217.

[12] Akaike, H. (1970b). A fundamental relation between predictor identification and power spectrum estimation, *Ann. Inst. Statist. Math.*, **22**, 219-223.

[13] Akaike, H. (1971). Autoregressive model fitting for control, *Ann. Inst. Statist. Math.*, **23**, 163-180.

[14] Akaike, H. (1972a). Use of an information theoretic quantity for statistical model identification, Proc. 5th Hawaii Inter. Conference on System Sciences, 249-250.

[15] Akaike, H. (1972b). Automatic data structure search by the maximum likelihood, Proc. 5th Hawaii Inter. Conference on System Sciences, 99-101.

[16] Akaike, H. (1973a). Maximum likelihood identification of Gaussian autoregressive moving average models, *Biometrika*, **60**, 255-265.

[17] Akaike, H. (1973b). Block Toeplitz matrix inversion, *SIAM J. Appl. Math.*, **24**, 234-241.

[18] Akaike, H. (1973c). Information theory and an extension of the maximum likelihood principle, in *2nd Inter. Symposium on Information Theory*, B. N. Petrov and F. Csáki, Eds., Akadémiai Kiadó, Budapest, 267-281.

[19] Akaike, H. (1974a). A new look at the statistical model identification, *IEEE Trans. Automatic Control*, **AC-19**, 716-723.

[20] Akaike, H. (1974b). Markovian representation of stochastic processes and its application to the analysis of autoregressive moving average processes, *Ann. Inst. Statist. Math.*, **26**, 363-387.

[21] Akaike, H. (1975). Markovian representation of stochastic processes by canonical variables, *SIAM J. Control*, **13**, 162-173.

[22] Akaike, H. (1976). Canonical correlation analysis of time series and the use of an information criterion, in *System Identification: Advances and Case Studies*, R. K. Mehra and D. G. Lainiotis, Eds., Academic Press, New York.

[23] Akaike, H. (1977a). On entropy maximization principle, in *Applications of Statistics*, P. R. Krishnaiah, Ed., North-Holland, Amsterdam, 27-41.

[24] Akaike, H. (1977b). An objective use of Bayesian models, *Ann. Inst. Statist. Math.*, **29**, 9-20.

[25] Akaike, H. (1978a). Time series analysis and control through parametric models, in *Applied Time Series Analysis*, D. F. Findley, Ed., Academic Press, New York, 1-23.

[26] Akaike, H. (1978b). A new look at the Bayes procedure, *Biometrika*, **65**, 53-59.

[27] Akaike, H. (1978c). A Bayesian analysis of the minimum AIC procedure, *Ann. Inst. Statist. Math.*, **30**, 9-14.

[28] Akaike, H. (1979). A Bayesian extension of the minimum AIC procedure of autoregressive model fitting, *Biometrika*, **66**, 237-242.

[29] Akaike, H. (1983). Statistical inference and measurement of entropy, *Scientific Inference, Data Analysis and Robustness*, G. E. P. Box, T. Leonard and C. F. Wu, Eds., Academic Press, New York, 165-189.

[30] Ali, M. M. (1977). Analysis of autoregressive-moving average models: estimation and prediction, *Biometrika*, **64**, 535-545. Corrections: **65** (1978), 677.

[31] An, H. Z. (1982). Estimation of prediction error variance, *Stochastic Processes and their Applications*, **13**, 39-43.

[32] An, H. Z. and Z. G. Chen (1986). The identification of ARMA model, in *Essays in Time Series and Applied Processes*; Special volume 23A of *J. Appl. Prob.*, J. Gani and M. B. Priestley, Eds., The Applied Probability Trust, Sheffield, 75-87.

[33] An, H. Z., Z. G. Chen and E. J. Hannan (1982). Autocorrelation, autoregression and autoregressive approximation, *Ann. Statist.*, **10**, 926-936.

[34] An, H. Z., Z. G. Chen and E. J. Hannan (1983). A note on ARMA estimation, *J. Time Series Anal.*, **4**, 9-17.

[35] Anděl, J. (1982). Fitting models in time series analysis, *Math. Operationasforsche. Statist., Ser. Statistics*, **13**, 121-143.

[36] Andersen, N. (1974). On the calculation of filter coefficients for maximum entropy spectral analysis, *Geophysics*, **39**, 69-72.

[37] Andersen, N. (1978). Comments on the performance of maximum entropy algorithms, *Proceedings of the IEEE*, **66**, 1581-1582.

[38] Anderson, O. D. (1975a). A note on differencing autoregressive moving average (p, q) processes, *Biometrika*, **62**, 521-523.

[39] Anderson, O. D. (1975b). The recursive nature of the stationarity and invertibility restraints on the parameters of mixed autoregressive-moving average processes, *Biometrika*, **62**, 704-706.

[40] Anderson, O. D. (1977a). A commentary on 'A survey of time series,' *Inter. Statistical Review*, **45**, 273-297.

[41] Anderson, O. D. (1977b). The time series concept of invertibility, *Math. Operationsforsch. Statist., Ser. Statistics*, **9**, 399-406.

[42] Anderson, O. D. (1978). On the invertibility conditions for moving average processes, *Math. Operationsforsch. Statist., Ser. Statistics*, **9**, 525-529.

[43] Anderson, O. D. (1980a). A new approach to ARMA modeling: Some comments, in *Analyzing Time Series*, O. D. Anderson, Ed., North-Holland, Amsterdam, 43-71.

[44] Anderson, O. D., Ed. (1980b). *Time Series*, North-Holland, Amsterdam.

[45] Anderson, O. D., Ed. (1980c). *Analyzing Time Series*, North-Holland, Amsterdam.

[46] Anderson, O. D., Ed. *Time Series Analysis: Theory and Practices*, **1** (1982a), **2** (1982b), **3** (1983), **4** (1984a), **5** (1984b), **6** (1985a), **7** (1985b), North-Holland, Amsterdam.

[47] Anderson, O. D. (1990). On the partial autocorrelations for an explosive process, *Commun. Statist.*, **A19**, 3505-3526.

[48] Anderson, O. D. and M. R. Perryman, Eds. (1981). *Time Series Analysis*, North-Holland, Amsterdam.

[49] Anderson, O. D. and M. R. Perryman, Eds. (1982). *Applied Time Series Analysis*, North-Holland, Amsterdam.

[50] Anderson, R. L. (1942). Distribution of the serial correlation coefficient, *Ann. Math. Statist.*, **13**, 1-13.

[51] Anderson, T. W. (1963). Determination of the order of dependence in normally distributed time series, in *Time Series Analysis*, M. Rosenblatt, Ed., Wiley, New York.

[52] Anderson, T. W. (1971). *The Statistical Analysis of Time Series*, Wiley, New York.

[53] Anderson, T. W. (1975). Maximum likelihood estimation of parameters of an autoregressive process with moving average residuals and other covariance matrices linear structure, *Ann. Statist.*, **3**, 1283-1304.

[54] Anderson, T. W. (1977). Estimation for autoregressive moving average models in the time and frequency domains, *Ann. Statist.*, **5**, 842-865.

[55] Anderson, T. W. (1984). *An Introduction to Multivariate Statistical Analysis (2nd Edition)*, Wiley, New York.

[56] Anderson, T. W. and R. P. Mentz (1980). On the structure of the likelihood function of autoregressive and moving average models, *J. Time Series Anal.*, **1**, 83-94.

[57] Anderson, T. W. and H. Rubin (1950). The asymptotic properties of estimates of the parameters of a single equation in a complete system of stochastic equations, *Ann. Math. Statist.*, **21**, 570-582.

[58] Anderson, T. W. and A. M. Walker (1964). On the asymptotic distribution of the autocorrelations of a sample from a linear stochastic process, *Ann. Math. Statist.*, **35**, 1296-1303.

[59] Ansley, C. F. (1979). An algorithm for the exact likelihood of a mixed autoregressive-moving average process, *Biometrika*, **66**, 59-65.

[60] Ansley, C. F. and R. Kohn (1983). Exact likelihood of vector autoregressive-moving average process with missing or aggregated data, *Biometrika*, **70**, 275-278.

[61] Ansley, C. F. and P. Newbold (1979). On the finite sample distribution of residual autocorrelations in autoregressive-moving average model, *Biometrika*, **66**, 547-553.

[62] Aoki, M. (1990). *State Space Modeling of Time Series (2nd Edition)*, Springer-Verlag, Berlin.

[63] Arcese, A. (1983). On the method of maximum entropy spectrum estimation, *IEEE Trans. Information Theory*, **IT-29**, 161-164.

[64] Ash, R. (1965). *Information Theory*, Interscience Publishers, New York.

[65] Åström, K. J. and P. Eykhoff (1971). System identification-A survey, *Automatika*, **7**, 123-162.

[66] Åström, K. J. and D. Q. Mayne (1983). A new algorithm for recursive estimation of controlled ARMA processes, in *Proc. 6th IFAC Symposium on Identification and System Parameters*, G. A. Bekey and G. W. Saridis, Eds., Pergamon, Oxford, 122-126.

[67] Åström, K. J. and T. Söderström (1974). Uniqueness of the maximum likelihood estimates of the parameters of an ARMA model, *IEEE Trans. Automatic Control*, **AC-19**, 769-773.

[68] Bahadur, R. R. and S. L. Zabell (1979). Large deviations of the sample mean in general vector spaces, *Ann. Prob.*, **7**, 587-621.

[69] Bai, Z. D., K. Subramanyam and L. C. Zhao (1988). On determination of the order of an autoregressive model, *J. Multivariate Anal.*, **27**, 40-52.

[70] Baillie, R. T. (1979a). The asymptotic mean squared error of multistep prediction from the regression model with autoregressive errors, *J. Amer. Statist. Assoc.*, **74**, 175-184.

[71] Baillie, R. T. (1979b). Asymptotic prediction mean squared error for vector autoregressive models, *Biometrika*, **66**, 675-678.

[72] Baker, G. A. (1975). *Essentials of Padé Approximants*, Academic Press, New York.

[73] Barndorff-Nielsen, O. and G. Schou (1973). On the parameterization of autoregressive model by partial correlations, *J. Multivariate Anal.*, **3**, 408-419.

[74] Barrodale, I. and R. E. Erickson (1980). Algorithms for least-squares linear prediction and maximum entropy spectral analysis - Part I: Theory and Part II: FORTRAN program, *Geophysics*, **45**, 420-446.

[75] Bartlett, M. S. (1946). On the theoretical specification and sampling properties of autocorrelated time series, *J. Roy. Statist. Soc. Suppl.*, **8**, 27-41, 85-97.

[76] Bartlett, M. S. and P. H. Diananda (1950). Extensions of Quenouille's test for autoregressive schemes, *J. Roy. Statist. Soc. Ser. B.*, **12**, 108-115.

[77] Bartlett, M. S. and D. V. Rajalakshman (1953). Goodness of fit tests for simultaneous autoregressive series, *J. Roy. Statist. Soc. Ser. B*, **15**, 107-124.

[78] Basawa, I. V., R. M. Huggins and R. G. Staudte (1985). Robust tests for time series with an application to first-order autoregressive processes, *Biometrika*, **72**, 559-571.

[79] Battaglia, F. (1983). Inverse autocovariances and a measure of linear determinism for a stationary process, *J. Time Series Anal.*, **4**, 79-87.

[80] Battaglia, F. (1986). Recursive estimation of the inverse correlation function, *Statistica*, **46**, 75-82.

[81] Battaglia, F. (1988). On the estimation of the inverse correlation function, *J. Time Series Anal.*, **9**, 1-10.

[82] Beamish, N. and M. B. Priestley (1981). A study of autoregressive and window spectral estimation, *Appl. Statist.*, **30**, 41-58.

[83] Bednar, J. B. and B. Roberts (1985). On the relationship between Levinson recursion and the R and S arrays for ARMA model identification, *Commun. Statist.*, **A14**, 1217-1248.

[84] Beguin, J. M., C. Gourieroux and A. Monfort (1980). Identification of a mixed autoregressive-moving average process: The corner method, in *Time Series*, O. D. Anderson, Ed., North-Holland, Amsterdam, 423-436.

[85] Bergland, G. D. (1969). A guided tour of the fast Fourier transform, *IEEE Spectrum*, **6**, 41-52.

[86] Berk, K. N. (1974). Consistent autoregressive spectral estimates, *Ann. Statist.*, **2**, 489-502.

[87] Berlekamp, E. R. (1968). *Algebraic Coding Theory*, McGraw-Hill, New York.

[88] Berndt, E. R. and N. E. Savin (1977). Conflict among criteria for testing hypotheses in the multivariate linear regression model, *Econometrica*, **45**, 1263-1278.

[89] Bessler, D. A. and J. K. Binkley (1980). Autoregressive filtering of some economic data using PRESS and FPE, *Proc. Busi. Econ. Statist., Sect., A.S.A.*, 261-265.

[90] Bhansali, R. J. (1973). A Monte Carlo comparison of the regression method and the spectral methods of prediction, *J. Amer. Statist. Assoc.*, **68**, 621-625.

[91] Bhansali, R. J. (1974). Asymptotic mean-square error of predicting more than one-step ahead using the regression method, *Appl. Statist.*, **23**, 35-42.

[92] Bhansali, R. J. (1978). Linear prediction by autoregressive model fitting in the time domain, *Ann. Statist.*, **6**, 224-231.

[93] Bhansali, R. J. (1980). Autoregressive and window estimates of the inverse correlation function, *Biometrika*, **67**, 551-566.

[94] Bhansali, R. J. (1983a). Estimation of the order of a moving average model from autoregressive and window estimates of the inverse correlation function, *J. Time Series Anal.*, **4**, 137-162.

[95] Bhansali, R. J. (1983b). The inverse partial correlation function of a time series and its applications, *J. Multivariate Anal.*, **13**, 310-327.

[96] Bhansali, R. J. (1983c). A simulation study of autoregressive and window estimators of the inverse correlation function, *Appl. Statist.*, **32**, 141-149.

[97] Bhansali, R. J. (1986). The criterion autoregressive transfer function of Parzen, *J. Time Series Anal.*, **7**, 79-104.

[98] Bhansali, R. J. (1988). Consistent order determination for processes with infinite variance, *J. Roy. Statist. Soc. Ser. B*, **50**, 46-60.

[99] Bhansali, R. J. (1990). On a relationship between the inverse of a stationary covariance matrix and the linear interpolater, *J. Appl. Prob*, **27**, 156-170.

[100] Bhansali, R. J. (1991). Consistent recursive estimation of the order of an autoregressive moving average process, *Inter. Statist. Review*, **59**, 81-96.

[101] Bhansali, R. J. and D. Y. Downham (1977). Some properties of the order of an autoregressive model selected by a generalization of Akaike's FPE criterion, *Biometrika*, **64**, 547-551.

[102] Bhargava, A. (1986). On the theory of testing for unit roots in observed time series, *Rev. Econ. Studies*, **53**, 369-384.

[103] Billingsley, P. (1961). *Statistical Inferences for Markov Processes*, University of Chicago Press, Chicago.

[104] Bittanti, S., Ed. (1989). *Time Series and Linear Systems; Lecture Notes in Control and Information Sciences, Vol. 86*, Springer-Verlag, Berlin.

[105] Blackman, R. G. and J. W. Tukey (1958). *The Measurement of Power Spectra from the Point of View of Communication Engineering*, Dover, New York.

[106] Blahut, R. (1985). *Fast Algorithms for Signal Processing*, Addison-Wesley, Reading, MA.

[107] Bloomfield, P. (1972). On the error of prediction of a time series, *Biometrika*, **59**, 501-507.

[108] Box, G. E. P. and G. M. Jenkins (1976). *Time Series Analysis, Forecasting and Control (Revised Edition)*, Holden-Day, San Francisco.

[109] Box, G. E. P., T. Leonard and C. F. Wu, Eds. (1983). *Scientific Inference, Data Analysis and Robustness*, Holden-Day, New York.

[110] Box, G. E. P. and D. A. Pierce (1970). Distribution of residual autocorrelations in autoregressive-integrated moving average time series models, *J. Amer. Statist. Assoc.*, **65**, 1509-1526.

[111] Box, G. E. P. and G. C. Tiao (1977). A canonical analysis of multiple time series, *Biometrika*, **64**, 355-365.

[112] Breiman, L. (1957). The individual ergodic theorem of information theory, *Ann. Math. Statist.*, **28**, 809-814. Correction: **31** (1960), 809-810.

[113] Breusch, T. S. (1978). Testing for autocorrelation in dynamic linear models, *Australian Economic Papers*, **17**, 334-355.

[114] Breusch, T. S. (1979). Conflict among criteria for testing hypotheses: Extensions and comments, *Econometrica*, **47**, 203-207.

[115] Breusch, T. S. and A. R. Pagan (1980). The Lagrange multiplier test and its applications to model specification in econometrics, *Rev. Econ. Studies*, **47**, 239-253.

[116] Brigham, E. O. (1974). *The Fast Fourier Transform*, Prentice-Hall, Englewood Cliffs, NJ.

[117] Brigham, E. O. and R. E. Morrow (1967). The fast Fourier transform, *IEEE Spectrum*, **4**, 63-70.

[118] Brillinger, D. R. (1981). *Time Series: Data Analysis and Theory*, Holden-Day, San Francisco.

[119] Brillinger, D. R. and P. R. Krishnaiah, Eds. (1983). *Handbook of Statistics, Vol. 3; Time Series in the Frequency Domain*, North-Holland, Amsterdam.

[120] Brillinger, D. R. and G. C. Tiao, Eds. (1980). *Directions in Time Series: Proceedings of the IMS Special Topics Meeting on Time Series Analysis at Iowa State University (1978)*, Institute of Mathematical Statistics, Hayward, CA.

[121] Brockwell, P. J. and R. A. Davis (1987). *Time Series: Theory and Methods*, Springer-Verlag, Berlin.

[122] Burg, J. P. (1967). Maximum entropy spectral analysis, in Proceedings of the 37th Meeting of the Society of Exploration Geophysics. Reprinted in *Modern Spectrum Analysis*, D. G. Childers, Ed., (1978), IEEE Press, New York, 34-41.

[123] Burg, J. P. (1968). A new analysis technique for time series data, presented at the NATO Advanced Study Institute Signal Processing with Emphasis on Underwater Acoustics, Enschede, The Netherlands. Reprinted in *Modern Spectrum Analysis*, D. G. Childers, Ed. (1978), IEEE Press, New York, 42-48.

[124] Burg, J. P. (1975). *Maximum entropy spectral analysis*, Ph.D. dissertation, Stanford University, Stanford.

[125] Bustos, O. H. and V. J. Yohai (1986). Robust estimates for ARMA models, *J. Amer. Statist. Soc.*, **81**, 155-168.

[126] Cadzow, J. A. (1982). Spectral estimation: An overdetermined rational model equation approach, *Proceedings of the IEEE*, **70**, 907-939.

[127] Caines, P. E. (1988). *Linear Stochastic Systems*, Wiley, New York.

[128] Cameron, M. A. (1978). The prediction variance and related statistics for stationary time series, *Biometrika*, **65**, 283-296.

[129] Capon, J. (1969). High resolution frequency-wavenumber spectrum analysis, *Proceedings of the IEEE*, **57**, 1408-1418.

[130] Cernuschi-Frías, B. and J. D. Rogers (1988). On the exact maximum likelihood estimation of Gaussian autoregressive processes, *IEEE Trans. Acoustics, Speech, Signal Processing*, **ASSP-36**, 922-924.

[131] Chan, N. H. (1988). The parameter inference for nearly nonstationary time series, *J. Amer. Statist. Assoc.*, **83**, 1050-1063.

[132] Chan, N. H. (1990). Inference for near-integrated time series with infinite variance, *J. Amer. Statist. Assoc.*, **85**, 1069-1082.

[133] Chan, N. H. and C. Z. Wei (1987). Asymptotic inference for nearly nonstationary AR(1) processes, *Ann. Statist.*, **15**, 1050-1063.

[134] Chan, N. H. and C. Z. Wei (1988). Limiting distributions of least squares estimates of unstable autoregressive processes, *Ann. Statist.*, **16**, 367-401.

[135] Chatfield, C. (1977). Some recent developments in time series analysis, *J. Roy. Statist. Soc. Ser. A*, **140**, 492-510.

[136] Chatfield, C. (1979). Inverse autocorrelations, *J. Roy. Statist. Soc. Ser. A.*, **142**, 363-377.

[137] Chen, C. H., Ed. (1989). *Applied Time Series Analysis*, World Scientific, Singapore.

[138] Chen, Z. G. (1985). The asymptotic efficiency of a linear procedure of estimation for ARMA models, *J. Time Series Anal.*, **6**, 53-62.

[139] Chernoff, H. (1952). A measure of asymptotic efficiency for tests of a hypothesis based on the sum of observations, *Ann. Math. Statist.*, **23**, 493-507.

[140] Childers, D. G., Ed. (1978). *Modern Spectrum Analysis*, IEEE Press, New York.

[141] Chitturi, R. V. (1974). Distribution of residual autocorrelations in multiple autoregressive schemes, *J. Amer. Statist. Assoc.*, **69**, 928-934.

[142] Chitturi, R. V. (1976). Distribution of multivariate white noise autocorrelations, *J. Amer. Statist. Assoc.*, **71**, 223-226.

[143] Choi, B. S. (1985). A conditional limit construction of the normal probability density, *Ann. Inst. Statist. Math.*, **37**, 535-539.

[144] Choi, B. S. (1986a). An algorithm for solving the extended Yule-Walker equations of an autoregressive moving-average time series, *IEEE Trans. Information Theory*, **IT-32**, 417-419.

[145] Choi, B. S. (1986b). On the relation between the maximum entropy probability density function and the autoregressive model, *IEEE Trans. Acoustics, Speech, Signal Processing*, **ASSP-34**, 1659-1661.

[146] Choi, B. S. (1987). A Newton-Raphson solution for MA parameters of mixed autoregressive moving-average process, *J. Korean Statist. Soc.*, **16**, 1-9.

[147] Choi, B. S. (1990a). An information theoretic spectral density, *IEEE Trans. Acoustics, Speech, Signal Processing*, **ASSP-38**, 717-721.

[148] Choi, B. S. (1990b). A derivation of the asymptotic distribution of the partial autocorrelation function of an autoregressive process, *Commun. Statist.*, **A19**, 547-553.

[149] Choi, B. S. (1990c). On the block LU decomposition of block Toeplitz matrices using the vector Yule-Walker equations, *Commun. Statist.*, **A19**, 2815-2827.

[150] Choi, B. S. (1990d). The asymptotic distribution of the extended Yule-Walker estimates of a mixed ARMA process, Technical Report 140, Department of Statistics, University of California, Santa Barbara, CA 93106, October 14, 1990.

[151] Choi, B. S. (1990e). The asymptotic joint distribution of the correlation determinant estimates of an ARMA process. Technical Report 147, Department of Statistics, University of California, Santa Barbara, CA 93106, November 6, 1990. Also, appear in *Commun. Statist.*, **A20** (1991), 2823-2835.

[152] Choi, B. S. (1990f). On the asymptotic distributions of the R and S arrays for identifying an ARMA process. Technical Report 149, Department of Statistics, University of California, Santa Barbara, CA 93106, November 15, 1990. Also, appear in *Commun. Statist.*, **A20**, (1991), 3187-3206.

[153] Choi, B. S. (1990g). The asymptotic distributions of the θ, λ and η functions for identifying a mixed ARMA process. Technical Report 152, Department of Statistics, University of California, Santa Barbara, CA 93106, November 27, 1990.

[154] Choi, B. S. (1990h). Two chi-square statistics for determining orders of an ARMA process. Technical Report 156, Department of Statistics, University of California, Santa Barbara, CA 93106, December 31, 1990.

[155] Choi, B. S. (1991a). On the asymptotic distribution of the generalized partial autocorrelation function in ARMA processes, *J. Time Series Anal.*, **12**, 193-205.

[156] Choi, B. S. (1991b). On the conditional probability density functions of multivariate uniform random vectors and multivariate normal random vectors, *J. Multivariate Anal.*, **38**, 241-244.

[157] Choi, B. S. (1991c). A time-domain interpretation of the KL spectrum, *IEEE Trans. Signal Processing*, **SP-39**, 721-723.

[158] Choi, B. S. (1991d). The Cholesky decomposition and the maximum entropy probability density of a stationary process. Technical Report 157, Department of Statistics, University of California, Santa Barbara, CA 93106, January 17, 1991.

[159] Choi, B. S. (1991e). Multivariate maximum entropy spectrum, Technical Report 159, Department of Statistics, University of California, Santa Barbara, CA 93106, January 23, 1991.

[160] Choi, B. S. (1991f). Conditional probability density characterization of Gaussian autoregressive processes, Technical Report 162, Department of Statistics, University of California, Santa Barbara, CA 93106, February 14, 1991.

[161] Choi, B. S. (1991g). ARMA model identification based on Quenouille-Walker's goodness of fit test statistic, Technical Report 168, Department of Statistics, University of California, Santa Barbara, CA 93106, March 30, 1991.

[162] Choi, B. S. (1991h). A probabilistic interpretation of the KL spectrum, Technical Report 177, Department of Statistics, University of California, Santa Barbara, CA 93106, June 20, 1991.

[163] Choi, B. S. (1991i). An algorithm for Hannan and Rissanen's ARMA modeling method, Technical Report 178, Department of Statistics, University of California, Santa Barbara, CA 93106, June 25, 1991.

[164] Choi, B. S. and T. M. Cover (1983). A conditional limit characterization of the maximum entropy spectral estimation, 1983 Inter. Symposium on Information Theory, IEEE Press, New York, 155-156.

[165] Choi, B. S. and T. M. Cover (1984). An information theoretic proof of Burg's maximum entropy spectrum, *Proceedings of the IEEE*, **72**, 1094-1095.

[166] Choi, B. S. and T. M. Cover (1987). A proof of Burg's theorem, in *Maximum-Entropy and Bayesian Spectral Analysis and Estimation Problems*, C. R. Smith and G. J. Erickson Eds., D. Reidel Pub. Co., Dordrecht, 75-84.

[167] Chow, G. C. (1981). A comparison of the information and posterior probability criteria for model selection, *J. Econometrics*, **16**, 21-33.

[168] Chow, J. C. (1972). On estimating the orders of an autoregressive moving-average process with uncertain observations, *IEEE Trans. Automatic Control*, **AC-17**, 707-709.

[169] Clarke, B. R. and E. J. Godolphin (1982). Comparative power studies for goodness of fit tests of time series models, *J. Time Series Anal.*, **3**, 141-151.

[170] Cleveland, W. S. (1972). The inverse autocorrelations of a time series and their applications, *Technometrics*, **14**, 277-298.

[171] Conte, S. D. and C. de Boor (1987). *Elementary Numerical Analysis (3rd Edition)*, McGraw-Hill, New York.

[172] Cooley, J. W. and J. W. Tukey (1965). An algorithm for the machine calculation of complex Fourier series, *Mathematics of Computations*, **19**, 297-301.

[173] Cooley, J. W., P. A. W. Lewis and P. D. Welch (1967). Historical notes on the fast Fourier transform, *IEEE Trans. Audio Electroacoustics*, **AU-15**, 76-79.

[174] Cooley, J. W., P. A. W. Lewis and P. D. Welch (1970a). The fast Fourier transform algorithm: Programming considerations in the calculation of sine, cosine and Laplace transforms, *J. Sound Vibration*, **12**, 315-337.

[175] Cooley, J. W., P. A. W. Lewis and P. D. Welch (1970b). The application of the fast Fourier transform algorithm to the estimation of spectra and cross-spectra, *J. Sound Vibration*, **12**, 339-352.

[176] Cooper, D. M. and R. Thompson (1977). A note on the estimation of the parameters of the autoregressive-moving average process, *Biometrika*, **64**, 625-628.

[177] Cooper, D. M. and E. F. Wood (1981). Estimation of the parameters of the Markovian representation of the autoregressive-moving average model, *Biometrika*, **68**, 320-322.

[178] Cooper, D. M. and E. F. Wood (1982). Indentifying multivariate time series models, *J. Time Series Anal.*, **3**, 153-164.

[179] Cover, T. M., B. S. Choi and I. T. Csiszár (1984). Markov conditioning involving large deviations leads to Markov chain suggested by maximum entropy principle, 1985 IEEE Information Theory Conference in Brighton, IEEE Press, New York.

[180] Cox, D. R. (1981). Statistical analysis of time series: Some recent developments, *Scand. J. Statist.*, **8**, 93-115.

[181] Cox, D. R. and D. V. Hinkley (1974). *Theoretical Statistics*, Chapman and Hall, London.

[182] Cox, D. R. and N. Reid (1987). Parameter orthogonality and approximate conditional inference, *J. Roy. Statist. Soc. Ser. B*, **49**, 1-39.

[183] Cressie, N. (1988). A graphical procedure for determining nonstationarity in time series, *J. Amer. Statist. Assoc.*, **83**, 1108-1116.

[184] Cryer, J. D. and J. Ledolter (1981). Small-sample properties of the maximum likelihood estimator in the first-order moving average model, *Biometrika*, **68**, 691-694.

[185] Csiszár, I. T., T. M. Cover and B. S. Choi (1987). Conditional limit theorems under Markov conditioning, *IEEE Trans. Information Theory*, **IT-33**, 788-801.

[186] Csiszár, I. T. and J. Körner (1981). *Information Theory : Coding Theorems for Discrete Memoryless Systems*, Akadémiai Kiadó, Budapest.

[187] Cybenko, G. (1979). Round-off error propagation in Durbin's, Levinson's and Trench's algorithms, *Proc. ICASSP 79*, 498-501.

[188] Cybenko, G. (1980). The numerical stability of the Levinson-Durbin algorithm for Toeplitz systems of equations, *SIAM J. Scientific Statist. Comput.*, **1**, 303-319.

[189] Cybenko, G. (1983). A general orthogonalization technique with applications to time series analysis and signal processing, *Mathematics of Computation*, **40**, 323-336.

[190] Cybenko, G. (1984). The numerical stability of the lattice algorithm for least squares linear prediction problems, *BIT*, **24**, 441-455.

[191] Cybenko, G. and C. Van Loan (1986). Computing the minimum eigenvalues of a symmetric positive definite Toeplitz matrix, *SIAM J. Scientific Statist. Comput.*, **7**, 123-131.

[192] Dahlhaus, R. and B. M. Pötscher (1989). Convergence results for maximum likelihood type estimators in multivariable ARMA models II, *J. Multivariate Anal.*, **30**, 241-244.

[193] Damsleth, E. and A. H. El-Shaarawi (1989). ARMA models with double-exponentially distributed noise, *J. Roy. Statist. Soc. Ser. B*, **51**, 61-69.

[194] Davies, N. and P. Newbold (1979). Some power studies of a portmanteau test of time series model specification, *Biometrika*, **66**, 153-155.

[195] Davies, N. and P. Newbold (1980). Forecasting with misspecified models, *Appl. Statist.*, **29**, 87-92.

[196] Davies, N. and J. D. Petruccelli (1984). On the use of the general partial autocorrelation function for order determination in ARMA(p, q) processes, *J. Amer. Statist. Assoc.*, **79**, 374-377.

[197] Davies, N., T. Spedding and W. Watson (1980). Autoregressive moving average processes with non-normal residuals, *J. Time Series Anal.*, **1**, 103-109.

[198] Davies, N., C. M. Triggs and P. Newbold (1977). Significance levels of the Box-Pierce portmanteau statistic in finite samples, *Biometrika*, **64**, 517-522.

[199] Davis, M. (1966). *A First Course in Functional Analysis*, Nelson, London.

[200] Davisson, L. D. (1965). The prediction error of stationary Gaussian time series of unknown covariance, *IEEE Trans. Information Theory*, **IT-11**, 527-532.

[201] Davisson, L. D. (1966). A theory of adaptive filtering, *IEEE Trans. Information Theory*, **IT-12**, 97-102.

[202] DeGroot, M. H. (1970). *Optimal Statistical Decision*, McGraw-Hill, New York.

[203] Deistler, M., W. Dunsmuir and E. J. Hannan (1978). Vector linear time series models: Corrections and extensions, *Adv. Appl. Prob.*, **10**, 360-372.

[204] Demeure, C. J. and L. L. Scharf (1987). Linear statistical model for stationary sequences and related algorithms for Cholesky factorization of Toeplitz matrices, *IEEE Trans. Acoustics, Speech, Signal Processing*, **ASSP-35**, 29-42.

[205] Denby, L. and D. Martin (1979). Robust estimation of the first-order autoregressive parameter, *J. Amer. Statist. Assoc.*, **74**, 140-146.

[206] Deuschel, J. D. and D. W. Stroock (1989). *Large Deviations*, Academic Press, Boston.

[207] Dickey, D. A. and W. A. Fuller (1979). Distribution of the estimators for autoregressive time series with a unit root, *J. Amer. Statist. Assoc.*, **74**, 427-431.

[208] Dickey, D. A. and W. A. Fuller (1981). Likelihood ratio statistics for autoregressive time series with a unit root, *Econometrica*, **49**, 1057-1072.

[209] Diggle, P. J. and S. L. Zeger (1989). A non-Gaussian model for time series with pulses, *J. Amer. Statist. Assoc.*, **84**, 354-359.

[210] Dixon, W. J. (1944). Further contributions to the problem of serial correlation, *Ann. Math. Statist.*, **25**, 631-650.

[211] Doob, J. L. (1953). *Stochastic Processes*, Wiley, New York.

[212] Dunsmuir, W. and E. J. Hannan (1976). Vector linear time series models, *Adv. Appl. Prob.*, **8**, 339-364.

[213] Duong, Q. P. (1984). On the choice of the order of autoregressive models: A ranking and selection approach, *J. Time Series Anal.*, **5**, 145-157.

[214] Durbin, J. (1959). Efficient estimation of parameters in moving average models, *Biometrika*, **46**, 306-316.

[215] Durbin, J. (1960a). The fitting of time series models, *Inter. Statistical Review*, **28**, 233-244.

[216] Durbin, J. (1960b). Estimation of parameters in time-series regression models, *J. Roy. Statist. Soc. Ser. B*, **22**, 139-153.

[217] Durbin, J. (1970). Testing for serial correlation in least-squares regression when some of the regressors are lagged dependent variables, *Econometrica*, **38**, 410-421.

[218] Durbin, J. and G. S. Watson (1950). Testing for serial correlation in least squares regression-I, *Biometrika*, **37**, 409-428.

[219] Durbin, J. and G. S. Watson (1951). Testing for serial correlation in least squares regression-II, *Biometrika*, **38**, 409-428.

[220] Durbin, J. and G. S. Watson (1971). Testing for serial correlation in least squares regression-III, *Biometrika*, **58**, 409-428.

[221] Edward, J. A. and M. M. Fitelson (1973). Notes on maximum entropy processing, *IEEE Trans. Information Theory*, **IT-19**, 232-234.

[222] El Gamal, A. and T. M. Cover (1980). Multiple user information theory, *Proceedings of the IEEE*, **68**, 1466-1483.

[223] Elliott, D. F. and K. R. Rao (1982). *Fast Transforms: Algorithms, Analysis, Applications*, Academic Press, New York.

[224] Ellis, R. S. (1985). *Entropy, Large Deviations and Statistical Mechanics*, Springer-Verlag, Berlin.

[225] Engle, R. F. (1984). Wald, likelihood ratio, and Lagrange multiplier tests in econometrics, in *Handbook of Econometrics, Vol. II*, Z. Griliches and M. D. Intriligator, Eds., North-Holland, Amsterdam, 775-826.

[226] Ensor, K. B. and H. J. Newton (1988). The effect of order estimation on estimating the peak frequency of an autoregressive spectral density, *Biometrika*, **75**, 587-589.

[227] Ensor, K. B. and H. J. Newton (1990). A recursive in order algorithm for least squares estimates of an autoregressive process, *J. Statist. Comput. Simul.*, **37**, 115-126.

[228] Evans, G. B. A. and N. E. Savin (1981). Testing for unit roots: 1, *Econometrica*, **49**, 753-779.

[229] Evans, G. B. A. and N. E. Savin (1984). Testing for unit roots: 2, *Econometrica*, **52**, 1241-1269.

[230] Faddeeva, V. N. (1959). *Computational Methods of Linear Algebra*, Dover, New York.

[231] Feder, M. and E. Weinstein (1984). On the finite maximum entropy extrapolation, *Proceedings of the IEEE*, **72**, 1660-1662.

[232] Findley, D. F., Ed. (1978). *Applied Time Series Analysis*, Academic Press, New York.

[233] Findley, D. F. (1980). Limiting autocorrelations and their uses in the identification of nonstationary ARMA models, *Bulletin of the Institute of Mathematical Statistics*, **7**, 293-294.

[234] Findley, D. F., Ed. (1981). *Applied Time Series Analysis II*, Academic Press, New York.

[235] Findley, D. F. (1984). On some ambiguities associated with the fitting of ARMA models to time series, *J. Time Series Anal.*, **5**, 213-225.

[236] Findley, D. F. (1985). On the unbiasedness property of AIC for exact or approximating linear stochastic time series models, *J. Time Series Anal.*, **6**, 229-252.

[237] Findley, D. F. (1986). The uniqueness of moving average representations with independent and identically distributed random variables for non-Gaussian stationary time series, *Biometrika*, **73**, 520-521.

[238] Fitts, J. (1973). Testing for autocorrelation in the autoregressive moving average error model, *J. Econometrics*, **1**, 363-376.

[239] Fotopoulos, S. B. and W. D. Ray (1983). Components of prediction errors for a stationary process with estimated parameters, *J. Time Series Anal.*, **4**, 1-8.

[240] Franke, J. (1985a). A Levinson-Durbin recursion for autoregressive-moving average processes, *Biometrika*, **72**, 573-581.

[241] Franke, J. (1985b). ARMA processes have maximal entropy among time series with prescribed autocovariances and impulse responses, *Adv. Appl. Prob.*, **17**, 810-840.

[242] Franke, J. , W. Härdle and D. Martin, Eds. (1984). *Robust and Nonlinear Time Series Analysis: Lecture Notes in Statistics, Vol. 26*, Springer-Verlag, New York.

[243] Friedlander, B. (1982a). Lattice filters for adaptive processing, *Proceedings of the IEEE*, **70**, 829-867.

[244] Friedlander, B. (1982b). Lattice methods for spectral estimation, *Proceedings of the IEEE*, **70**, 990-1017.

[245] Fuhrmann, D. R. and B. Liu (1986). Rotational search methods for adaptive Pisarenko harmonic retrieval, *IEEE Trans. Acoustics, Speech, Signal Processing*, **ASSP-34**, 1550-1565.

[246] Fuller, W. A. (1976). *Introduction to Statistical Time Series*, Wiley, New York.

[247] Fuller, W. A. (1985). Nonstationary autoregressive time series, in *Handbooks of Statistics, Vol. 5*, E. J. Hannan, P. R. Krishnaiah and M. M. Rao, Eds., North-Holland, Amsterdam, 1-23.

[248] Fuller, W. A. and D. P. Hasza (1981). Properties of predictors for autoregressive time series, *J.Amer. Statist. Assoc.*, **76**, 155-161.

[249] Fuller, W. A., D. P. Hasza and J. J. Goebel (1981). Estimation of the parameters of stochastic difference equations, *Ann. Statist.*, **9**, 531-543.

[250] Gabr, M. M. and T. Subba Rao (1981). The estimation and prediction of subset bilinear time series models with applications, *J. Time Series Anal.*, **2**, 155-171.

[251] Gani, J. and M. B. Priestley, Eds. (1986). *Essays in Time Series and Allied Process: Papers in Honor of E. J. Hannan*, Special volume 23A of *J. Applied Probability*, The University Press, Belfast.

[252] Geisser, S. and W. F. Eddy (1979). A prediction approach to model selection, *J. Amer. Statist. Assoc.*, **74**, 153-160.

[253] Gentleman, W. M. and G. Sande (1966). Fast Fourier Transform - for fun and profits, *Proc. AFIPS 1966 Fall Joint Computer Conference*, **28**, 563-578.

[254] Gersch, W. and G. Kitagawa (1983). The prediction of time series with trends and seasonalities, *J. Business Econ. Statist.*, **1**, 253-264.

[255] Gersch, W. and D. R. Sharpe (1973). Estimation of power spectra with finite-order autoregressive model, *IEEE Trans. Automatic Control*, **AC-18**, 367-369.

[256] Glasbey, C. A. (1982). A generalization of partial autocorrelations useful in identifying ARMA models, *Technometrics*, **24**, 223-228.

[257] Godfrey, L. G. (1978a). Testing against general autoregressive and moving average error models when the regressors include lagged dependent variables, *Econometrica*, **46**, 1293-1301.

[258] Godfrey, L. G. (1978b). Testing for higher order serial correlation in regression equations when the regressors include lagged dependent variables, *Econometrica*, **46**, 1303-1310.

[259] Godfrey, L. G. (1979). Testing the adequacy of a time series model, *Biometrika*, **66**, 67-72.

[260] Godolphin, E. J. (1977). A direct representation for the maximum likelihood estimator of a Gaussian moving average process, *Biometrika*, **64**, 375-384.

[261] Godolphin, E. J. (1978). Modified maximum likelihood estimation of Gaussian moving averages using a pseudoquadratic convergence criterion, *Biometrika*, **65**, 203-206.

[262] Godolphin, E. J. (1980a). An invariance property for the maximum likelihood estimator of the parameters of a Gaussian moving average process, *Ann. Statist.*, **8**, 1093-1099.

[263] Godolphin, E. J. (1980b). A method for testing the order of an autoregressive-moving average process, *Biometrika*, **67**, 699-703.

[264] Godolphin, E. J. (1984). A direct representation for the large-sample maximum likelihood estimator of a Gaussian autoregressive-moving average process, *Biometrika*, **71**, 281-289.

[265] Godolphin, E. J. and J. G. de Gooijer (1982). On the maximum likelihood estimation of the parameters of a Gaussian moving average process, *Biometrika*, **69**, 443-451.

[266] Godolphin, E. J. and J. M. Unwin (1983). Evaluation of the covariance matrix for the maximum likelihood estimator of a Gaussian autoregressive moving average process, *Biometrika*, **70**, 279-284.

[267] Golub, G. H. and C. F. Van Loan (1989). *Matrix Computation (2nd Edition)*, The Johns Hopkins University Press, Baltimore.

[268] Gooijer, J. G. de, B. Abraham, A. Gould and L. Robinson (1985). Methods for determining the order of an autoregressive-moving average process: a survey, *Inter. Statistical Review*, **53**, 301-329.

[269] Gooijer, J. G. de and R. M. J. Heuts (1981). The corner method: an investigation of an order determination procedure for general ARMA processes, *J. Oper. Research Soc.*, **32**, 1039-1046.

[270] Gooijer, J. G. de and P. Saikkonen (1988). A specification strategy for order determination in ARMA models, *Commun. Statist.*, **B17**, 1037-1054.

[271] Grandell, J., M. Hamrud, and P. Toll (1980). A remark on the correspondence between the maximum entropy method and the autoregressive model, *IEEE Trans. Information Theory*, **IT-26**, 750-751.

[272] Granger, C. W. J. (1979). Nearer-normality and some econometric models, *Econometrica*, **47**, 781-784.

[273] Granger, C. W. J. and A. Andersen (1978). On the invertibility of time series models, *Stochastic Processes and their Applications*, **8**, 87-92.

[274] Graupe, D., D. J. Krause and J. B. Moore (1975). Identification of autoregressive moving-average parameters of time series, *IEEE Trans. Automatic Control*, **AC-20**, 104-107.

[275] Gray, H. L., A. G. Houston and F. W. Morgan (1978). On G- spectral estimation, *Applied Time Series Analysis*, D. F. Findley, Ed., Academic Press, New York, 39-138.

[276] Gray, H. L., G. D. Kelley and D. D. McIntire (1978). A new approach to ARMA modeling, *Commun. Statist.*, **B7**, 1-77.

[277] Gray, H. L. and W. R. Schucany (1972). *The Generalized Jackknife Statistic*, Marcel Dekker, New York.

[278] Gray, H. L. and W. A. Woodward (1981). Application of S-arrays to seasonal data, *Applied Time Series Analysis II*, D. F. Findley, Ed., Academic Press, New York, 379-413.

[279] Grenander, U. and M. Rosenblatt (1957). *Statistical Analysis of Stationary Time Series*, Wiley, New York.

[280] Grenander, U. and G. Szegö (1955). *Toeplitz Forms and their Applications*, Chelsea Pub. Co., New York.

[281] Gutowski, P. R., E. A. Robinson and S. Treitel (1978). Spectral estimation: Fact or fiction, *IEEE Trans. Geosci. Electron.*, **GE-16**, 80-84.

[282] Haggan, V. and O. B. Oyetunji (1984). On the selection of subset autoregressive time series models, *J. Time Series Anal.*, **5**, 103-113.

[283] Hall, A. (1989). Testing for a unit root in the premence of moving average errors, *Biometrika*, **76**, 49-56.

[284] Hallin, M. (1980). Invertibility and generalized invertibility of time series models, *J. Roy. Statist. Soc. Ser. B*, **42**, 210-212.

[285] Hallin, M. (1981). Addendum to "Invertibility and generalized invertibility of time series models," *J. Roy. Statist. Soc. Ser. B*, **43**, 103.

[286] Hallin, M. (1984). Spectral factorization of nonstationary moving average processes, *Ann. Statist.*, **12**, 172-192.

[287] Hallin, M. (1986). Non-stationary $q-$dependent processes and time-varying moving-average models: Invertibility properties and the forecasting problem, *Adv. Appl. Prob.*, **18**, 170-210.

[288] Hallin, M. and J. Ingenbleek (1983). Nonstationary Yule-Walker equations, *Statistics and Probability Letters*, **1**, 189-195.

[289] Hallin, M., C. Lefevre and M. L. Puri (1988). On time-reversibility and the uniqueness of moving average representations for non-Gaussian stationary time series, *Biometrika*, **75**, 170-171.

[290] Hamilton, D. C. and D. G. Watts (1978). Interpreting partial autocorrelation functions of seasonal time series models, *Biometrika*, **65**, 135-140.

[291] Hannan, E. J. (1970a). *Multiple Time Series*, Wiley, New York.

[292] Hannan, E. J. (1970b). The seasonal adjustment of economic time series, *Inter. Econ. Rev.*, **11**, 1-29.

[293] Hannan, E. J. (1976). The asymptotic distribution of serial covariances, *Ann. Statist.*, **4**, 396-399.

[294] Hannan, E. J. (1979). A note on autoregressive-moving average identification, *Biometrika*, **66**, 672-674.

[295] Hannan, E. J. (1980a). Recursive estimation based on ARMA models, *Ann. Statist.*, **8**, 762-777. Correction: **9** (1981), 233.

[296] Hannan, E. J. (1980b). The estimation of the order of an ARMA process, *Ann. Statist.*, **8**, 1071-1081.

[297] Hannan, E. J. (1981). Estimating the dimension of a linear system, *J. Multivariate Anal.*, **11**, 459-473.

[298] Hannan, E. J. (1982). Testing for autocorrelation and Akaike's criterion, in *Essays in Statistical Science*, Special volume 19A of *J. Appl. Prob.*, The Applied Probability Trust, Sheffield, 403-412.

[299] Hannan, E. J. and M. Deistler (1988). *The Statistical Theory of Linear Systems*, Wiley, New York.

[300] Hannan, E. J., W. T. M. Dunsmuir and M. Deistler (1980). Estimation of vector ARMAX models, *J. Multivariate Anal.*, **10**, 275-295.

[301] Hannan, E. J. and C. C. Heyde (1972). On limit theorems for quadratic functions of discrete time series, *Ann. Math. Statist.*, **43**, 2058-2066.

[302] Hannan, E. J. and L. Kavalieris (1983a). Linear estimation of ARMA processes, *Automatica*, **19**, 447-448.

[303] Hannan, E. J. and L. Kavalieris (1983b). The convergence of autocorrelations and autoregressions, *Austral. J. Statist.*, **25**, 287-297.

[304] Hannan, E. J. and L. Kavalieris (1984a). A method for autoregressive-moving average estimation, *Biometrika*, **72**, 273-280.

[305] Hannan, E. J. and L. Kavalieris (1984b). Multivariate linear time series models, *Adv. Appl. Prob.*, **16**, 492-561.

[306] Hannan, E. J. and L. Kavalieris (1986a). Regression, autoregression models, *J. Time Series Anal.*, **7**, 27-49.

[307] Hannan, E. J. and L. Kavalieris (1986b). The convergence of autocorrelations and autoregressions, *Austral. J. Statist.*, **25**, 287-297.

[308] Hannan, E. J., L. Kavalieris and M. Mackisack (1986). Recursive estimation of linear systems, *Biometrika*, **73**, 119-133.

[309] Hannan, E. J., P. R. Krishnaiah and M. M. Rao, Eds. (1985). *Handbook of Statistics, Vol. 5: Time Series in the Time Domain*, North-Holland, Amsterdam.

[310] Hannan, E. J. and A. J. McDougall (1988). Regression procedures for ARMA estimation, *J. Amer. Statist. Assoc.*, **83**, 490-498.

[311] Hannan, E. J., A. J. McDougall and D. S. Poskitt (1989). Recursive estimation of autoregressions, *J. Roy. Statist. Soc. Ser. B*, **51**, 217-233.

[312] Hannan, E. J. and D. S. Poskitt (1988). Unit canonical correlations between future and past, *Ann. Statist.*, **16**, 784-790.

[313] Hannan, E. J. and B. G. Quinn (1979). The determination of the order of an autoregression, *J. Roy. Statist. Soc. Ser. B*, **41**, 190-195.

[314] Hannan, E. J. and J. Rissanen (1982). Recursive estimation of mixed autoregressive-moving average order, *Biometrika*, **69**, 81-94. Correction: **70** (1983), 303.

[315] Harris, B., Ed. (1967). *Spectral Analysis of Time Series*, Wiley, New York.

[316] Harris, F. J. (1978). On the use of windows for harmonic analysis with the discrete Fourier transform, *Proceeding of the IEEE*, **66**, 51-83.

[317] Harvey, A. C. (1989). *Forecasting, Structural Time Series Models and the Kalman Filter*, Cambridge University Press, Cambridge.

[318] Harvey, A. C. and G. D. A. Phillips (1979). Maximum likelihood estimation of regression models with autoregressive-moving average disturbances, *Biometrika*, **66**, 49-58.

[319] Hastings, C. (1955). *Approximations for Digital Computers*, Princeton University Press, Princeton.

[320] Hasza, D. P. (1980). The asymptotic distribution of the sample autocorrelations for an integrated ARMA process, *J. Amer. Statist. Assoc.*, **75**, 349-352.

[321] Hasza, D. P. and Fuller, W. A. (1979). Estimation for autoregressive processes with unit roots, *Ann. Statist.*, **7**, 1106-1120.

[322] Haykin, S., Ed. (1979). *Nonlinear Methods of Spectral Analysis: Topics in Applied Physics, Vol. 34*, Springer-Verlag, Berlin.

[323] Haykin, S. and J. A. Cadzow, Eds. (1982). *Proceedings of the IEEE; Special Issue on Spectral Estimation*, **70**, IEEE Press, New York.

[324] Haykin, S. and S. Kesler (1979). Prediction-error filtering and maximum-entropy spectral estimation, in *Nonlinear Methods of Spectral Analysis*, S. Haykin, Ed., Springer-Verlag, New York, 9-72.

[325] Heyde, C. C. and D. J. Scott (1973). Invariance principle for the law of the iterated logarithm for martingales and processes with stationary increments, *Ann. Prob.*, **1**, 428-436.

[326] Hipel, K. W., A. I. McLeod and W. C. Lennox (1977). Advances in Box-Jenkins modelling, 1. Mode construction, *Water Resources Research*, **13**, 567-575.

[327] Hjorth, U. and L. Holmqvist (1981). On model selection based on validation with applications to pressure and temperature prognosis, *Appl. Statist.*, **30**, 264-274.

[328] Hocking, R. R. and R. N. Leslie (1967). Selection of the best subset in regression analysis, *Technometrics*, **9**, 531-540.

[329] Hokstad, P. (1983). A method for diagnostic checking of time series models, *J. Time Series Anal.*, **4**, 177-183.

[330] Hopwood, W. S. and P. Newbold (1980). Time series analysis in accounting: A survey and analysis of recent issues, *J. Time Series Anal.*, **1**, 135-144.

[331] Hosking, J. R. M. (1978). A unified derivation of the asymptotic distributions of goodness-of-fit statistics for autoregressive time-series models, *J. Roy. Statist. Soc. Ser. B*, **40**, 341-349.

[332] Hosking, J. R. M. (1980a). The asymptotic distribution of the sample inverse autocorrelations of an autoregressive-moving average process, *Biometrika*, **67**, 223-226.

[333] Hosking, J. R. M. (1980b). Lagrange-multiplier tests of time-series models, *J. Roy. Statist. Soc. Ser. B*, **42**, 170-181.

[334] Hosking, J. R. M. (1980c). The multivariate portmanteau statistic, *J. Amer. Statist. Assoc.*, **75**, 602-607.

[335] Hosking, J. R. M. (1981a). Lagrange-multiplier tests of multivariate time-series models, *J. Roy. Statist. Soc. Ser. B*, **43**, 219-230.

[336] Hosking, J. R. M. (1981b). Equivalent forms of the multivariate portmanteau statistic, *J. Roy. Statist. Soc. Ser. B*, **43**, 261-262. Correction: **51** (1989), 303.

[337] Hsiao, C. (1979). Causality tests in econometrics, *J. Economic Dynamics and Control*, **1**, 321-346.

[338] Huang, D. (1990a). Selecting order for general autoregressive models by minimum description length, *J. Time Series Anal.*, **11**, 107-119.

[339] Huang, D. (1990b). Levinson-type recursive algorithms for least-squares autoregression, *J. Time Series Anal.*, **11**, 295-315.

[340] Hurvich, C. M. and K. I. Beltrão (1990). Cross-validatory choice of a spectrum estimate and its connections with AIC, *J. Time Series Anal.*, **11**, 121-137.

[341] Hurvich, C. M. and C. L. Tsai (1989). Regression and time series model selection in small samples, *Biometrika*, **76**, 297-307.

[342] Hurvich, C. M. and S. L. Zeger (1990). A frequency domain seclection criterion for regression with autocorrelated errors, *J. Amer. Statist. Assoc.*, **85**, 705-714.

[343] Huzii, M. (1981). Estimation of coefficients of an autoregressive process by using a higher order moment, *J. Time Series Anal.*, **2**, 87-93.

[344] IBM Corporation (1968). *IBM System/360 Scientific Subroutine Package*, IBM System Reference Library, New York.

[345] Ihara, S. (1984). Maximum entropy spectral analysis and ARMA processes, *IEEE Trans. Information Theory*, **IT-30**, 377-380.

[346] Janacek, G. (1975). Estimation of the minimum mean square error of prediction, *Biometrika*, **62**, 175-180.

[347] Jaynes, E. T. (1957a). Information theory and statistical mechanics, *Physical Review*, **106**, 620-630.

[348] Jaynes, E. T. (1957b). Information theory and statistical mechanics II, *Physical Review*, **108**, 171-190.

[349] Jaynes, E. T. (1968). Prior probabilities, *IEEE Trans. Systems Science and Cybernetics*, **SSC-4**, 227-241.

[350] Jaynes, E. T. (1978). "Where do we stand on maximum entropy?", in *The Maximum Entropy Formalism*, R. D. Levine and M. Tribus, Eds., (1978), MIT Press, Cambridge, MA, 15-118.

[351] Jaynes, E. T. (1982). On the rationale of maximum-entropy methods, *Proceedings of the IEEE*, **70**, 939-952.

[352] Jenkins, G. M. (1965). A survey of spectral analysis, *Appl. Statist.*, **14**, 2-32.

[353] Jenkins, G. M. and A. S. Alavi (1981). Some aspects of modelling and forecasting multivariate time series, *J. Time Series Anal.*, **2**, 1-47.

[354] Jenkins, G. M. and L. G. Watts (1968). *Spectral Analysis and its Applications*, Academic Press, New York.

[355] Jones, R. H. (1974). Identification and autoregressive spectrum estimations, *IEEE Trans. Automatic Control*, **AC-19**, 894-897.

[356] Jones, R. H. (1975). Fitting autoregressions, *J. Amer. Statist. Assoc.*, **70**, 590-592.

[357] Jones, R. H. (1976). Estimation of the innovation generalized variance of a multivariate stationary time series, *J. Amer. Statist. Assoc.*, **71**, 386-388.

[358] Jong, P., de (1976). The recursive fitting of autoregressions, *Biometrika*, **63**, 525-530.

[359] Jong, P., de (1988). A cross-validation filter for time series models, *Biometrika*, **75**, 594-600.

[360] Kabaila, P. V. (1980). An optimality property of the least-squares estimate of the parameter of the spectrum of a purely nondeterministic time series, *Ann. Statist.*, **8**, 1082-1092.

[361] Kabaila, P. V. (1983). On estimating time series parameters using sample autocorrelations, *J. Roy. Statist. Soc. Ser. B*, **45**, 107-119.

[362] Kabaila, P. V. (1987). On Rissanen's lower bound on the accumulated mean-square prediction error, *J. Time Series Anal.*, **8**, 301-309.

[363] Kagan, A. M., Y. V. Linnik and C. R. Rao (1973). *Characterization Problems in Mathematical Statistics*, Wiley, New York.

[364] Kailath, T. (1974). A view of three decades of linear filtering theory, *IEEE Trans. Information Theory*, **IT-20**, 146-181.

[365] Kalman, R. E. (1960). A new approach to linear filtering and prediction theory problems, *Trans. Amer. Soc. Mech. Eng., J. Basic Engineering*, **82**, 35-45.

[366] Kalman, R. E. (1963). Mathematical description of linear dynamical systems, *SIAM J. Control*, **1**, 152-192.

[367] Kalman, R. E. and R. S. Bucy (1961). New results in linear filtering and prediction theory, *Trans. Amer. Soc. Mech. Eng., J. Basic Engineering*, **83**, 95-108.

[368] Kanto, A. J. (1984). A characterization of the inverse autocorrelation function, *Commun. Statist.*, **A13**, 2503-2510.

[369] Kanto, A. J. (1987). A formula for the inverse autocorrelation function of an autoregressive process, *J. Time Series Anal.*, **8**, 311-312.

[370] Kashyap, R. L. (1977). A Bayesian comparison of different classes of dynamic models using empirical data, *IEEE Trans. Automatic Control*, **AC-22**, 715-727.

[371] Kashyap, R. L. (1978). Optimal feature selection and decision rules in classification problems with time series, *IEEE Trans. Information Theory*, **IT-24**, 281-288.

[372] Kashyap, R. L. (1980). Inconsistency of the AIC rule for estimating the order of autoregressive models, *IEEE Trans. Automatic Control*, **AC-25**, 996-998.

[373] Kashyap, R. L. (1982). Optimal choice of AR and MA parts in autoregressive moving-average models, *IEEE Trans. Pattern Analysis and Machine Intelligence*, **PAMI-4**, 99-104.

[374] Kashyap, R. L. and R. E. Nasburg (1974). Parameter estimation in multivariate stochastic difference equations, *IEEE Trans. Automatic Control*, **AC-19**, 784-797.

[375] Kavalieris, L. (1989). The estimation of the order of an autoregression using recursive residuals and cross-validation, *J. Time Series Anal.*, **10**, 271-281.

[376] Kaveh, M. and G. R. Cooper (1976). An empirical investigation of the properties of the autoregressive spectral estimator, *IEEE Trans. Information Theory*, **IT-22**, 313-323.

[377] Kawashima, H. (1980). Parameter estimation of autoregressive integrated processes by least squares, *Ann. Statist.*, **8**, 423-435.

[378] Kay, S. M. (1983). Recursive maximum likelihood estimation of autoregressive processes, *IEEE Trans. Acoustics, Speech, and Signal Processing*, **ASSP-31**, 56-65.

[379] Kay S. M. and S. L. Marple, Jr. (1981). Spectrum analysis - A modern perspective, *Proceedings of the IEEE*, **69**, 1380-1419.

[380] Kay S. M. and A. K. Shaw (1988). Frequency estimation by principal component AR spectral estimation method without eigendecomposition, *IEEE Trans. Acoustics, Speech, Signal Processing*, **ASSP-36**, 95-101.

[381] Kedem, B. and E. Slud (1981). On goodness of fit of time series models: An application of high order crossings, *Biometrika*, **68**, 551-556.

[382] Kendall, M. G. (1954). Note on bias in the estimation of autoregression, *Biometrika*, **41**, 403-404.

[383] Kendall, M. G. and A. Stuart (1979). *The Advanced Theory of Statistics, Vol. 2 (4th Edition)*, Griffin, London.

[384] Kendall, M. G., A. Stuart and J. K. Ord (1983). *The Advanced Theory of Statistics, Vol. 3 (4th Edition)*, Griffin, London.

[385] Kennedy, W. J. and J. E. Gentle (1980). *Statistical Computing*, Marcel Dekker, New York.

[386] Kesler, S. B., Ed. (1986). *Modern Spectrum Analysis II*, IEEE Press, New York.

[387] Khinchin, A. I. (1956). On the fundamental theorems of information theory, *Usp. Mat. Nauk*, **11**, 17-75. English translation is in *Mathematical Foundations of Information Theory*, R. A. Silverman and M. D. Friedman, Translators, (1975), Dover, New York, 17-75.

[388] Kitagawa, G. (1977). On a search procedure for the optimal ARMA order, *Ann. Inst. Statist. Math.*, **29**, 319-332.

[389] Kitagawa, G. (1981). A nonstationary time series model and its fitting by a recursive filter, *J. Time Series Anal.*, **2**, 103-116.

[390] Knoke, J. D. (1975). Testing for randomness against autocorrelated alternatives: The parametric case, *Biometrika*, **62**, 571-575.

[391] Knoke, J. D. (1977). Testing for randomness against autocorrelation: Alternative tests, *Biometrika*, **64**, 523-529.

[392] Kohn, R. (1977). Note concerning the Akaike and Hannan estimation procedures for an autoregressive-moving average process, *Biometrika*, **64**, 622-625.

[393] Konvalinka, I. S. and M. Matausek (1979). Simultaneous estimation of poles and zeros in speech analysis and ITIF-iterative inverse filtering algorithm, *IEEE Trans. Acoustics, Speech, Signal Processing*, **ASSP-27**, 485-492.

[394] Koreisha, S. G. and T. M. Pukkila (1987). Identification of nonzero elements in the polynomial matrices of mixed VARMA processes, *J. Roy. Statist. Soc. Ser. B*, **49**, 112-126.

[395] Koreisha, S. G. and T. M. Pukkila (1988). Estimation of the polynomial matrices of vector moving average processes, *J. Statist. Comput. Simul.*, **28**, 313-343.

[396] Koreisha, S. G. and T. M. Pukkila (1990a). A generalized least-squares approach for estimation of autoregressive moving-average models, *J. Time Series Anal.*, **11**, 139-151.

[397] Koreisha, S. G. and T. M. Pukkila (1990b). Linear methods for estimating ARMA and regression models with serial correlation, *Commun. Statist.*, **B19**, 71-102.

[398] Koreisha, S. G. and G. Yoshimoto (1991). A comparison among identification procedures for autoregressive moving average models, *Inter. Statist. Review*, **59**, 37-57.

[399] Kozin, F. and F. Nakajima (1980). The order determination problem for linear time-varying AR models, *IEEE Trans. Automatic Control*, **AC-25**, 250-257.

[400] Kromer, R. E. (1969). Asymptotic properties of the autoregressive spectral estimators, Ph.D. Dissertation, Stanford University, Stanford.

[401] Kullback, S. (1959). *Information Theory and Statistics*, Dover, New York.

[402] Kullback, S. and R. A. Leibler (1951). On information and sufficiency, *Ann. Math. Statist.*, **22**, 79-86.

[403] Kulperger, R. (1985). On an optimality property of Whittle's Gaussian estimate of the parameter of the spectrum of a time series, *J. Time Series Anal.*, **6**, 253-259.

[404] Kunitomo, N. and T. Yamamoto (1985). Properties of predictors in misspecified autoregressive time series models, *J. Amer. Statist. Assoc.*, **80**, 941-950.

[405] Lacoss, R. T. (1971). Data adaptive spectral analysis methods, *Geophysics*, **36**, 661-675.

[406] Larimore, W. E. (1983). Predictive inference, sufficiency, entropy and an asymptotic likelihood principle, *Biometrika*, **70**, 175-181.

[407] Lauritzen, S. L. (1981). Time series analysis in 1880: A discussion of contributions made by T. N. Thiele, *Inter. Statist. Rev.*, **49**, 319-331.

[408] Lawrance, A. J. and P. A. W. Lewis (1980). The exponential autoregressive-moving average EARMA(p, q) process, *J. Roy. Statist. Soc. Ser. B*, **42**, 150-161.

[409] Leamer, E. E. (1979). Information criteria for choice of regression models: A comment, *Econometrica*, **47**, 507-510.

[410] Ledolter, J. and B. Abraham (1981). Parsimony and its importance in time series forecasting, *Technometrics*, **23**, 411-414.

[411] Lee, R. C. K. (1964). *Optimal Estimation Identification and Control*, The MIT Press, Cambridge.

[412] Lee, D. T. L., B. Friedlander and M. Morf (1982). Recursive ladder algorithms for ARMA modeling, *IEEE Trans. Automatic Control*, **AC-27**, 753-764.

[413] Lee, D. T. L., M. Morf and B. Friedlander (1981). Recursive least squares ladder estimation algorithms, *IEEE Trans. Acoustics, Speech, Signal Processing*, **ASSP-29**, 627-641.

[414] Lehmann, E. L. (1959). *Testing Statistical Hypotheses*, Wiley, New York.

[415] Levinson, N. (1947). The Wiener RMS (Root Mean Square) error criterion filter design and prediction, *J. Mathematics and Physics*, **25**, 261-278.

[416] Li, W. K. (1985). Distribution of residual autocorrelations in multivariate autoregressive index models, *Biometrika*, **72**, 686-688.

[417] Li, W. K. and Y. V. Hui (1989). Robust multiple time series modelling, *Biometrika*, **76**, 309-315.

[418] Li, W. K. and A. I. McLeod (1981). Distribution of the residual autocorrelations in multivariate ARMA time series models, *J. Roy. Statist. Soc. Ser. B*, **43**, 231-239.

[419] Lii, K. S. (1985). Transfer function model order and parameter estimation, *J. Time Series Anal.*, **6**, 153-169.

[420] Lindberger, N. A. (1973). Comments on 'On estimating the orders of an autoregressive moving-average process with uncertain observations,' *IEEE Trans. Automatic Control*, **AC-18**, 689-691.

[421] Liu, L. M. and D. M. Hanssens (1982). Identification of multiple-input transfer function models, *Commun. Statist.*, **A11**, 297-314.

[422] Ljung, G. M. (1982). The likelihood function for a stationary Gaussian autoregressive-moving average process with missing observations, *Biometrika*, **69**, 265-268.

[423] Ljung, G. M. (1986). Diagnostic testing of univariate time series models, *Biometrika*, **73**, 725-730.

[424] Ljung, G. M. (1988). On the Lagrange multiplier test for autoregressive moving-average models, *J. Time Series Anal.*, **9**, 355-359.

[425] Ljung, G. M. and G. E. P. Box (1978). On a measure of lack of fit in time series models, *Biometrika*, **65**, 297-303.

[426] Ljung, G. M. and G. E. P. Box (1979). The likelihood function of stationary autoregressive-moving average models, *Biometrika*, **66**, 265-270.

[427] Lütkepohl, H. (1985). Comparison of criteria for estimating the order of a vector autoregressive process, *J. Time Series Anal.*, **6**, 35-52.

[428] Lysne, D. and D. Tjøstheim (1987). Loss of spectral peaks in autoregressive spectral estimation, *Biometrika*, **74**, 200-206.

[429] McClave, J. T. (1973). On the bias of autoregressive approximations to moving averages, *Biometrika*, **60**, 599-605.

[430] McClave, J. T. (1975). Subset autoregression, *Technometrics*, **17**, 213-220.

[431] McClave, J. T. (1978a). Estimating the order of autoregressive models: The max χ^2 method, *J. Amer. Statist. Assoc.*, **73**, 122-128.

[432] McClave, J. T. (1978b). Estimating the order of moving average models: The max χ^2 method, *Commun. Statist.*, **A7**, 259-276.

[433] McDonough, R. N. (1979). Application of the maximum-likelihood method and the maximum entropy method to array processing, in *Nonlinear Methods of Spectral Analysis*, S. Haykin, Ed., Springer-Verlag, New York, 181-244.

[434] McDunnough, P. (1979). The estimation of an autocorrelation parameter of a Gaussian vector process, *J. Roy. Statist. Soc. Ser. B*, **41**, 203-209.

[435] McLeod, A. I. (1975). Derivation of the theoretical autocovariance function of autoregressive-moving average time series, *Appl. Statist.*, **24**, 255-256. Correction: **26** (1977), 194.

[436] McLeod, A. I. (1978). On the distribution of residual autocorrelations in Box-Jenkins models, *J. Roy. Statist. Soc. Ser. B*, **40**, 296-302.

[437] McLeod, A. I. (1984). Duality and other properties of multiplicative seasonal autoregressive-moving average models, *Biometrika*, **71**, 207-211.

[438] McLeod, A. I., K. W. Hipel and W. C. Lennox (1977). Advances in Box-Jenkins modelling, 2. Applications, *Water Resources Research*, **13**, 577-586.

[439] McLeod, A. I. and W. K. Li (1983). Diagnostic checking ARMA time series models using squared-residual autocorrelations, *J. Time Series Anal.*, **4**, 269-273.

[440] McMillian, B. (1953). The basic theorems of information theory, *Ann. Math. Statist.*, **24**, 196-219.

[441] Makhoul, J. (1981). On the eigenvectors of symmetric Toeplitz matrices, *IEEE Trans. Acoustics, Speech, Signal Processing*, **ASSP-29**, 868-872.

[442] Makridakis, S. (1976). A survey of time series, *Inter. Statistical Review*, **44**, 29-70.

[443] Makridakis, S. (1978). Time-series analysis and forecasting: an update and evaluation, *Inter. Statistical Review*, **46**, 255-278.

[444] Makridakis, S. and S. C. Wheelwright, Eds. (1979). *Forecasting: Studies in the Management Sciences, Vol. 12 of TIMS*, North-Holland, Amsterdam.

[445] Mallows, C. L. (1973). Some comments on C_p, *Technometrics*, **15**, 661-675.

[446] Mandrekar, V. and H. Salehi, Eds. (1983). *Prediction Theory and Harmonic Analysis*, North-Holland, Amsterdam.

[447] Mann, H. B. and A. Wald (1943). On the statistical treatment of linear stochastic difference equations, *Econometrica*, **11**, 173-220.

[448] Marple, S. L. Jr. (1980). A new autoregressive spectrum analysis algorithm, *IEEE Trans. Acoustics, Speech, Signal Processing*, **ASSP-28**, 441-454.

[449] Marriott, F. H. C. and J. A. Pope (1954). Bias in the estimation of autocorrelations, *Biometrika*, **41**, 390-402.

[450] Martin, R. D. (1980). Robust estimation of autoregressive models, in *Directions in Time Series*, D. R. Brillinger and G. C. Tiao, Eds., Institute of Mathematical Statistics, Hayward, CA., 228-262.

[451] Martin, R. D. (1981). Robust methods for time series, *Applied Time Series II*, D. F. Findley, Ed., Academic Press, New York, 683-759.

[452] Martin, R. D. and V. J. Yohai (1985). Robustness in time series and estimating ARMA models, in *Handbook of Statistics, Vol. 5*, E. J. Hannan, P. R. Krishnaiah and M. M. Rao, Eds., North-Holland, Amsterdam, 119-155.

[453] Masarotto, G. (1987). Robust and consistent estimates of autoregressive-moving average parameters, *Biometrika*, **74**, 791-797.

[454] Massey, J. L. (1969). Shift-register synthesis and BCH coding, *IEEE Trans. Information Theory*, **IT-15**, 122-127.

[455] Mayne, D. Q. and F. Firoozan (1982). Linear identification of ARMA processes, *Automatica*, **18**, 461-466.

[456] Mehra, R. K. and D. G. Lainiotis, Eds. (1976). *System Identification*, Academic Press, New York.

[457] Mentz, R. P. (1977). Estimation in the first order moving average model through the finite autoregressive approximation, *J. Econometrics*, **6**, 225-236.

[458] Merikoski, J. K. and T. M. Pukkila (1983). A note on the expectation of products of autocorrelations, *Biometrika*, **70**, 528-529.

[459] Milhøj, A. (1981). A test of fit in time series models, *Biometrika*, **68**, 177-187.

[460] Mittnik, S. (1990). Computation of theoretical autocovariance matrices of multivariate autoregressive moving average time series, *J. Roy. Statist. Soc. Ser. B*, **52**, 151-155.

[461] Monahan, J. F. (1984). A note on enforcing stationarity in autoregressive-moving average models, *Biometrika*, **71**, 403-404.

[462] Moran, P. A. P. (1948). Some theorems on time series: II. The significance of the serial correlation coefficient, *Biometrika*, **35**, 255-260.

[463] Morettin, P. A. (1984). The Levinson algorithm and its applications in time series analysis, *Inter. Statistical Review*, **52**, 83-92.

[464] Morf, M., A. Vieira and T. Kailath (1978). Covariance characterization by partial autocorrelation matrices, *Ann. Statist.*, **6**, 643-648.

[465] Neave, H. R. (1972). A comparison of lag window generators, *J. Amer. Statist. Assoc.*, **67**, 152-158.

[466] Neftci, S. N. (1982). Specification of economic time series models using Akaike's criterion, *J. Amer. Statist. Assoc.*, **77**, 537-540.

[467] Nelson, C. R. (1974). The first-order moving average process, *J. Econometrics*, **2**, 121-141.

[468] Newbold, P. (1974). The exact likelihood function for a mixed autoregressive-moving average process, *Biometrika*, **61**, 423-426.

[469] Newbold, P. (1980). The equivalence of two tests of time series model adequacy, *Biometrika*, **67**, 463-465.

[470] Newbold, P. (1981). Some recent developments in time series analysis, *Inter. Statistical Review*, **49**, 53-66.

[471] Newbold, P. (1984). Some recent developments in time series analysis-II, *Inter. Statistical Review*, **52**, 53-66.

[472] Newbold, P. (1988). Some recent developments in time series analysis-III, *Inter. Statistical Review*, **56**, 53-66.

[473] Newbold, P. and T. Bos (1983). On q-conditioned partial correlations, *J. Time Series Anal.*, **4**, 53-55.

[474] Newton, H. J. and M. Pagano (1983). The finite memory prediction of covariance stationary time series, *SIAM J. Scient. Statist. Comput.*, **4**, 320-339.

[475] Nicholls, D. F. (1972). On Hannan's estimation of ARMA models, *Austral. J. Statist.*, **3**, 262-269.

[476] Nicholls, D. F. (1973). Frequency domain estimation procedures for linear models, *Biometrika*, **60**, 202-205.

[477] Nicholls, D. F. (1976). The efficient estimation of vector linear time series models, *Biometrika*, **63**, 381-390.

[478] Nicholls, D. F. (1977). A comparison of estimation methods for vector linear time series models, *Biometrika*, **64**, 85-90.

[479] Nicholls, D. F. and A. D. Hall (1979). The exact likelihood function of multivariate autoregressive-moving average models, *Biometrika*, **66**, 259-264.

[480] Nicholls, D. F. and A. R. Pagan (1985). Varying coefficient regression, in *Handbook of Statistics, Vol. 5*, E. J. Hannan, P. R. Krishnaiah and M. M. Rao, Eds., North-Holland, Amsterdam, 413-449.

[481] Nicholls, D. F. and A. L. Pope (1988). Bias in the estimation of multivariate autoregressions, *Austral. J. Statist.*, **30A**, 296-309.

[482] Nitzberg, R. (1979). Spectral estimation: An impossibility?, *Proceedings of the IEEE*, **67**, 437-439.

[483] O'Brien, P. C. (1980). A likelihood test for multivariate serial correlation, *Biometrika*, **67**, 531-537.

[484] Ogata, Y. (1980). Maximum likelihood estimation of incorrect Markov models for time series and the derivation of AIC, *J. Appl. Prob.*, **17**, 59-72.

[485] Osborn, D. R. (1976). Maximum likelihood estimation of moving average processes, *Ann. Econom. Social Measurement*, **5**, 75-87.

[486] Osborn, D. R. (1977). Exact and approximate maximum likelihood estimators for vector moving average processes, *J. Roy. Statist. Soc. Ser. B*, **39**, 114-118.

[487] Ozaki, T. (1977). On the order determination of ARIMA models, *Appl. Statist.*, **26**, 290-301.

[488] Pagano, M. (1972). An algorithm for fitting autoregressive scheme, *Appl. Statist.*, **21**, 274-281.

[489] Pagano, M. (1973). When is an autoregressive scheme stationary?, *Commun. Statist.*, **A1**, 533-544.

[490] Pagano, M. (1974). Estimation of models of autoregressive signal plus white noise, *Ann. Statist.*, **2**, 99-108.

[491] Parzen, E. (1967). *Time Series Analysis Papers*, Holden-Day, San Francisco.

[492] Parzen, E. (1968). Statistical spectral analysis (Single channel case) in 1968, Technical Report No. 11, Dept. of Statistics, Stanford University, Stanford.

[493] Parzen, E. (1969). Multiple time series modelling, in *Multivariate Analysis - II*, P. R. Krishnaiah, Ed., Academic Press, New York, 389-409.

[494] Parzen, E. (1971). Efficient estimation of a stationary time series mixed schemes, *Bull. Inst. Inter. Statist.*, **44**, 315-319.

[495] Parzen, E. (1974). Some recent advances in time series modelling, *IEEE Trans. Automatic Control*, **AC-19**, 389-409.

[496] Parzen, E. (1975). Some solution to the time series modelling and prediction problem, in *The Search for Oil*, D. Owen, Ed., Marcel Dekker, New York, 1-16.

[497] Parzen, E. (1977). Multiple time series: Determining the order of approximating autoregressive schemes, *Multivariate Analysis IV*, P. R. Krishnaiah, Ed., North-Holland, Amsterdam, 283-295.

[498] Parzen, E. (1979a). Forecasting and whitening filter estimation, *Management Sciences*, **12**, 149-165.

[499] Parzen, E. (1979b). Nonparametric statistical data modeling, *J. Amer. Statist. Assoc.*, **74**, 105-131.

[500] Parzen, E. (1980a). Time series modeling, spectral analysis and forecasting, in *Directions in Time Series*, D. R. Brillinger and G. C. Tiao, Eds., Institute of Mathematical Statistics, Hayward, CA., 80-111.

[501] Parzen, E. (1980b). Multiple time series modelling, *Multivariate Analysis V*, P. R. Krishnaiah, Ed., Academic Press, New York, 389-409.

[502] Parzen, E. (1983a). Autoregressive spectral estimation, *Handbook of Statistics, Vol. 3*, D. R. Brillinger and P. R. Krishnaiah, Eds., North-Holland, Amsterdam, 221-247.

[503] Parzen, E., Ed. (1983b). *Time Series Analysis of Irregularly Observed Data: Lecture Notes in Statistics, Vol. 25*, Springer-Verlag, Berlin.

[504] Parzen, E. (1986). Quantile spectral analysis and long-memory time series, in *Essays in Time Series and Applied Processes*; Special volume 23A of *J. Appl. Prob.*, J. Gani and M. B. Priestley, Eds., The Applied Probability Trust, Sheffield, 41-54.

[505] Parzen, E. and H. J. Newton (1980). Multiple time series modeling II, *Multivariate Analysis V*, P. R. Krishnaiah, Ed., North-Holland, Amsterdam, 181-197.

[506] Parzen, E. and M. Pagano (1979). An approach to modeling seasonality stationary time series, *J. Econometrics*, **9**, 137-153.

[507] Paulsen, J. (1984). Order determination of multivariate autoregressive time series with unit roots, *J. Time Series Anal.*, **5**, 115-127.

[508] Paulsen, J. and D. Tjøstheim (1985). On the estimation of residual variance and order in autoregressive time series, *J. Roy. Statist. Soc. Ser. B*, **47**, 216-228.

[509] Pearlman, J. G. (1980). An algorithm for the exact likelihood of a high-order autoregressive-moving average process, *Biometrika*, **67**, 232-233.

[510] Penm, J. H. W. and R. D. Terrell (1982). On the recursive fitting of subset autoregressions, *J. Time Series Anal.*, **3**, 43-59.

[511] Perron, P. (1989). The calculation of the limiting distribution of the least-squares estimator in a near-integrated model, *Econometric Theory*, **5**, 241-255.

[512] Petruccelli, J. D. and N. Davies (1984). Some restrictions on the use of corner method hypothesis tests, *Commun. Statist.*, **A13**, 543-551.

[513] Petruccelli, J. D. and N. Davies (1986). A portmanteau test for self-exciting threshold autoregressive-type nonlinearlity in time series, *Biometrika*, **73**, 687-694.

[514] Petrushev, P. P. and V. A. Popov (1987). *Rational Approximation of Real Functions*, Cambridge University Press, Cambridge.

[515] Phadke, M. S. and G. Kedem (1978). Computation of the exact likelihood function of multivariate moving average models, *Biometrika*, **65**, 511-519.

[516] Pham, D. T. (1978). On the fitting of multivariate processes of the autoregressive-moving average type, *Biometrika*, **65**, 99-107.

[517] Pham, D. T. (1979). The estimation of parameters for autoregressive-moving average models from sample autocovariances, *Biometrika*, **66**, 555-560.

[518] Pham, D. T. (1984a). The estimation of parameters for autoregressive moving average models, *J. Time Series Anal.*, **5**, 53-68.

[519] Pham, D. T. (1984b). A note on some statistics useful in identifying the order of autoregressive moving average model, *J. Time Series Anal.*, **5**, 273-279.

[520] Pham, D. T. (1986). A frequency domain approach to Lagrange multiplier test for autoregressive moving average models, *J. Time Series Anal.*, **7**, 73-78.

[521] Pham, D. T. (1987). Exact maximum likelihood estimate and Lagrange multiplier test statistic for ARMA models, *J. Time Series Anal.*, **8**, 61-78.

[522] Pham, D. T. (1988). Estimation of autoregressive parameters and order selection for ARMA models, *J. Time Series Anal.*, **9**, 265-279.

[523] Phillips, P. C. B. (1987a). Time series regression with a unit root, *Econometrica*, **55**, 277-301.

[524] Phillips, P. C. B. (1987b). Towards a unified asymptotic theory for autoregression, *Biometrika*, **74**, 535-547.

[525] Phillips, P. C. B. and P. Perron (1988). Testing for a unit root in time series regression, *Biometrika*, **75**, 335-346.

[526] Piccolo, D. (1982). The size of the stationarity and invertibility region of an autoregressive-moving average process, *J. Time Series Anal.*, **3**, 245-247.

[527] Piccolo, D. and G. Tunnicliffe-Wilson (1984). A unified approach to ARMA model identification and preliminary estimation, *J. Time Series Anal.*, **5**, 183-204.

[528] Pierce, D. A. (1985). Testing normality in autoregressive models, *Biometrika*, **72**, 293-297.

[529] Pino, F. A., P. A. Morettin and R. P. Mentz (1987). Modelling and forecasting linear combinations of time series, *Inter. Statistical Review*, **55**, 295-313.

[530] Pisarenko, V. F. (1972). On the estimation of spectra by means of nonlinear functions of the covariance matrix, *Geophys. J. Roy. Astronom. Soc.*, **28**, 511-531.

[531] Pisarenko, V. F. (1973). The retrieval of harmonics from a covariance function, *Geophys. J. Roy. Astronom. Soc.*, **33**, 347-366.

[532] Pope, A. L. (1990). Biases of estimators in multivariate non-Gaussian autoregressions, *J. Time Series Anal.*, **11**, 249-258.

[533] Porat, B. (1986). On estimation of the parameters of vector Gaussian processes from sample covariances, Proc. 25th Conf. Decision and Control, Athens, IEEE Press, New York, 2002-2005.

[534] Porat, B. (1987). Some asymptotic properties of the sample covariances of Gaussian autoregressive moving-average processes, *J. Time Series Anal.*, **8**, 205-220.

[535] Porat, B. and B. Friedlander (1984). An efficient algorithm for output-error model reduction, *Inter. J. Control*, **39**, 95-113.

[536] Porter-Hudak, S. (1990). An application of the seasonal fractionally differenced model to the monetary aggregates, *J. Amer. Statist. Assoc.*, **85**, 338-344.

[537] Poskitt, D. S. (1987). A modified Hannan-Rissanen strategy for mixed autoregressive-moving average order determination, *Biometrika*, **74**, 781-790.

[538] Poskitt, D. S. (1989). A method for the estimation and identification of transfer function models, *J. Roy. Statist. Soc. Ser. B*, **51**, 29-46. Correction: **52** (1990), 391.

[539] Poskitt, D. S. and A. R. Tremayne (1980). Testing the specification of a fitted autoregressive-moving average model, *Biometrika*, **67**, 359-363.

[540] Poskitt, D. S. and A. R. Tremayne (1981a). A time series application of the use of Monte Carlo methods to compare statistical tests, *J. Time Series Anal.*, **2**, 263-277.

[541] Poskitt, D. S. and A. R. Tremayne (1981b). An approach to testing linear time series models, *Ann. Statist.*, **9**, 974-986.

[542] Poskitt, D. S. and A. R. Tremayne (1982). Diagnostic tests for multiple time series models, *Ann. Statist.*, **10**, 114-120.

[543] Poskitt, D. S. and A. R. Tremayne (1983). On the posterior odds of time series models, *Biometrika*, **70**, 157-162.

[544] Poskitt, D. S. and A. R. Tremayne (1986). Some aspects of the performance of diagnostic checks in bivariate time series models, *J. Time Series Anal.*, **7**, 217-233.

[545] Poskitt, D. S. and A. R. Tremayne (1987). Determining a portfolio of linear time series models, *Biometrika*, **74**, 125-137.

[546] Pötscher, B. M. (1983). Order estimation in ARMA-models by Lagrangian multiplier tests, *Ann. Statist.*, **11**, 872-885.

[547] Pötscher, B. M. (1985). The behavior of the Lagrangian multiplier test in testing the orders of an ARMA-model, *Metrika*, **32**, 129-150.

[548] Pötscher, B. M. (1987). Convergence results for maximum likelihood type estimators in multivariable ARMA models, *J. Multivariate Anal.*, **21**, 29-52.

[549] Pötscher, B. M. (1990). Estimation of autoregressive moving-average order given an infinite number of models and approximation of spectral densities, *J. Time Series Anal.*, **11**, 165-179.

[550] Priestley, M. B. (1981). *Spectral Analysis and Time Series (2 Vols.)*, Academic Press, London.

[551] Priestley, M. B. (1988). *Non-linear and Non-stationary Time Series Analysis*, Academic Press, London.

[552] Prothero, D. L. and K. F. Wallis (1976). Modelling macroeconomic time series, *J. Roy. Statist. Soc. Ser. A*, **139**, 468-500.

[553] Pukkila, T. M. (1982). On the identification of ARMA(p, q) models, in *Time Series Analysis: Theory and Practice I*, O. D. Anderson, Ed., North-Holland, Amsterdam, 81-103.

[554] Pukkila, T. M. (1988). An improved estimation method for univariate autoregressive models, *J. Multivariate Anal.*, **27**, 422-433.

[555] Pukkila, T. M. (1989). Joint determination of autoregressive order and degree of differencing, in *Statistical Data Analysis and Inference*, Y. Dodge, Ed., North-Holland, Amsterdam, 519-531.

[556] Pukkila, T. M., S. Koreisha and A. Kallinen (1990). The identification of ARMA models, *Biometrika*, **77**, 537-548.

[557] Pukkila, T. M. and P. R. Krishnaiah (1988). On the use of autoregressive order determination criteria in univariate white noise tests, *IEEE Trans. Acoustics, Speech, Signal Processing*, **ASSP-36**, 764-774.

[558] Pye, W. C. and H. L. Atchison (1973). An algorithm for the computation of the higher order G-transformation, *SIAM J. Numer. Anal.*, **10**, 1-7.

[559] Quenouille, M. H. (1947). A large-sample test for the goodness of fit of autoregressive schemes, *J. Roy. Statist. Soc. Ser. A*, **110**, 123-129.

[560] Quenouille, M. H. (1949a). The joint distribution of serial correlation coefficients, *Ann. Math. Statist.*, **20**, 561-571.

[561] Quenouille, M. H. (1949b). Approximate tests of correlation in time series, *J. Roy. Statist. Soc. Ser. B*, **11**, 68-84.

[562] Quinn, B. G. (1980a). Limiting behaviour of autocorrelation function of ARMA process as several roots of characteristic equation approach unit circle, *Commun. Statist.*, **B9**, 195-198.

[563] Quinn, B. G. (1980b). Order determination for a multivariate autoregression, *J. Roy. Statist. Soc. Ser. B*, **42**, 182-185.

[564] Quinn, B. G. (1986). Testing for the presence of sinusoidal components, in *Essays in Time Series and Applied Processes*; Special volume 23A of *J. Appl. Prob.*, J. Gani and M. B. Priestley, Eds., The Applied Probability Trust, Sheffield, 201-210.

[565] Quinn, B. G. (1988). A note on AIC order determination for multivariate autoregressions, *J. Time Series Anal.*, **9**, 241-245.

[566] Quinn, B. G. (1989). Estimating the number of terms in a sinusoidal regression, *J. Time Series Anal.*, **10**, 71-75.

[567] Ramasubban, T. A. (1972). An approximate distribution of a noncircular serial correlation coefficient, *Biometrika*, **59**, 79-84.

[568] Ramsey, F. L. (1974). Characterization of the partial autocorrelation function, *Ann. Statist.*, **2**, 1296-1301.

[569] Rao, C. R. (1947). Large sample tests of statistical hypotheses concerning several parameters with applications to problems of estimation, *Proceedings of the Cambridge Philosophical Society*, **44**, 50-57.

[570] Rao, C. R. (1973). *Linear Statistical Inference and Its Applications (2nd Edition)*, Wiley, New York.

[571] Rao, M. M. (1961). Consistency and limit distributions of estimators of parameters in explosive stochastic difference equation, *Ann. Math. Statist.*, **32**, 195-218.

[572] Reeves, J. E. (1972). The distribution of the maximum likelihood estimator of the parameter in the first-order autoregressive series, *Biometrika*, **59**, 387-394.

[573] Reinsel, G. (1980). Asymptotic properties of prediction errors for the multivariate autoregressive model using estimated parameters, *J. Roy. Statist. Soc. Ser. B*, **42**, 328-333.

[574] Reinsel, G. (1983). Some results on multivariate autoregressive index models, *Biometrika*, **70**, 145-156.

[575] Rezayat, F. and G. Anandalingam (1988). Using instrumental variables for selecting the order of ARMA models, *Commun. Statist.*, **A17**, 3029-3065.

[576] Rissanen, J. (1973). Algorithms for triangular decomposition of block Hankel and Toeplitz matrices with application to factoring positive matrix polynomials, *Mathematics of Computation*, **27**, 147-154.

[577] Rissanen, J. (1978). Modeling by shortest data description, *Automatica*, **14**, 465-471.

[578] Rissanen, J. (1979). Shortest data description and consistency of order estimates in ARMA processes, in *Inter. Symposium on Systems Optimization and Analysis*, A. Bensoussan and J. L. Lions, Eds., Springer-Verlag, Berlin, 92-98.

[579] Rissanen, J. (1980). Consistent order estimates of autoregressive processes by shortest description of data, in *Analysis and Optimization of Stochastic Systems*, O. L. R. Jacobs et. al. Eds., Academic Press, New York, 451-461.

[580] Rissanen, J. (1982). Estimation of structure by minimum description length, *Circuits Systems Signal Process*, **1**, 395-406.

[581] Rissanen, J. (1983a). A universal prior for integers and estimation by minimum description length, *Ann. Statist.*, **11**, 416-431.

[582] Rissanen, J. (1983b). Order estimation in Box-Jenkins model for time series, *Methods of Operations Research*, **44**, 143-150.

[583] Rissanen, J. (1984). Universal coding, information, prediction and estimation, *IEEE Trans. Information Theory*, **IT-30**, 629-636.

[584] Rissanen, J. (1986a). Stochastic complexity and modeling, *Ann. Statist.*, **14**, 1080-1100.

[585] Rissanen, J. (1986b). Order estimation by accumulated prediction errors, in *Essays in Time Series and Applied Processes*; Special volume 23A of *J. Appl. Prob.*, J. Gani and M. B. Priestley, Eds., The Applied Probability Trust, Sheffield, 55-61.

[586] Rissanen, J. (1987). Stochastic complexity, *J. Roy. Statist. Soc. Ser. B*, **49**, 223-265.

[587] Rissanen, J. (1989). *Stochastic Complexity in Statistical Inquiry*, World Scientific, Singapore.

[588] Robinson, E. A. (1982). A historical perspective of spectrum estimation, *Proceedings of the IEEE*, **70**, 885-907.

[589] Robinson, P. M. (1983). Review of various approaches to power spectrum estimation, in *Handbook of Statistics, Vol. 3*, D. R. Brillinger and P. R. Krishnaiah, Eds., North-Holland, Amsterdam, 383-368.

[590] Roebuck, P. A. and S. Barnett (1978). A survey of Toeplitz and related matrices, *Int. J. Systems Sci.*, **9**, 921-934.

[591] Rosenblatt, M. Ed. (1963). *Proceedings of the Symposium on Time Series Analysis, Brown University, June 11-14, 1962*, Wiley, New York.

[592] Roy, R. (1977). On the asymptotic behaviour of the sample autocovariance function for an integrated moving average process, *Biometrika*, **64**, 419-421.

[593] Roy, R. (1989). Asymptotic covariance structure of serial correlations in multivariate time series, *Biometrika*, **76**, 824-827.

[594] Rudra, A. (1952). Discrimination in time-series analysis, *Biometrika*, **39**, 434-439.

[595] Rudra, A. (1954). A critical survey of some test methods in time series analysis, *Calcutta Statistical Association Bulletin*, **5**, 165-177.

[596] Runge, C. (1903) Uber die Zerlegung empirisch gegebener periodischer Funktionen in Sinuswellen, *Z. Math. Phys.*, **48**, 443-456.

[597] Said, S. E. and D. A. Dickey (1984). Testing for unit roots in autoregressive-moving average models of unknown order, *Biometrika*, **71**, 599-607.

[598] Said, S. E. and D. A. Dickey (1985). Hypothesis testing in ARIMA$(p,1,q)$ models, *J. Amer. Statist. Assoc.*, **80**, 369-374.

[599] Saikkonen, P. (1983). Asymptotic relative efficiency of some tests of fit in time series models, *J. Time Series Anal.*, **4**, 69-78.

[600] Saikkonen, P. (1986). Asymptotic properties of some preliminary estimates for autoregressive moving average time series models, *J. Time Series Anal.*, **7**, 133-155.

[601] Sakai, H. (1981). Asymptotic distribution of the order selected by AIC in multivariate autoregressive model fitting, *Inter. J. Control*, **33**, 175-180.

[602] Sanov, I. N. (1957). On the probability of large deviations of random variables, in *IMS and AMS Selected Translations in Mathematical Statistics and Probability, vol. 1*, 1961, 213-244.

[603] Sargan, D. and A. Bhargava (1983). Testing for residuals from least squares regression for being generated by the Gaussian random walk, *Econometrica*, **51**, 151-174.

[604] Savin, N. E. (1976). Conflict among testing procedures in a linear regression model with autoregressive disturbances, *Econometrica*, **44**, 1303-1313.

[605] Sawa, T. (1978). Information criteria for discriminating among alternative regression models, *Econometrica*, **46**, 1273-1291.

[606] Schmidt, P. (1974). The asymptotic distribution of forecasts in the dynamic simulation of an econometric model, *Biometrika*, **63**, 303-309.

[607] Schuster, A. (1898). On the investigation of hidden periodicities with application to a supposed 26-day period of meteological phenomena, *Terrestrial Magnetism*, **3**, 13-41.

[608] Schuster, A. (1899). The periodogram of magnetic declination as obtained from the records of the Greenwich observatory during the years 1871 - 1895, *Transactions of the Cambridge Philosophical Society*, **XVIII**, 107-135.

[609] Schuster, A. (1906a). The periodogram and its optical analogy, *Proceedings of the Royal Society of London*, **77**, 136-140.

[610] Schuster, A. (1906b). On the periodicities of sunspots, *Philosophical Transactions of the Royal Society of London: Ser. A*, **206**, 69-100.

[611] Schwarz, G. (1971). A sequential student test, *Ann. Math. Statist.*, **42**, 1003-1009.

[612] Schwarz, G. (1978). Estimating the dimension of a model, *Ann. Statist.*, **6**, 461-464.

[613] Schwert, G. W. (1989). Tests for unit roots, *J. Business Econ. Statist.*, **7**, 147-160.

[614] Shaman, P. (1975). An approximate inverse for the covariance matrix of moving average and autoregressive processes, *Ann. Statist.*, **3**, 532-538.

[615] Shaman, P. (1976). Approximations for stationary covariance matrices and their inverses with application to ARMA models, *Ann. Statist.*, **4**, 292-301.

[616] Shaman, P. and R. A. Stine (1988). The bias of autoregressive coefficient estimators, *J. Amer. Statist. Assoc.*, **83**, 842-848.

[617] Shannon, C. E. (1948). A mathematical theory of communication, *Bell System Technical Journal*, **27**, 379-423. Reprinted in *The Mathematical Theory of Information* (1949), University of Illinois Press, Urbana.

[618] Shea, B. L. (1987). Estimation of multivariate time series, *J. Time Series Anal.*, **8**, 95-109.

[619] Shenton, L. R. and W. L. Johnson (1965). Moments of a serial correlation coefficients, *J. Econometrics*, **27**, 308-320.

[620] Shibata, R. (1976). Selection of the order of an autoregressive model by Akaike's information criterion, *Biometrika*, **63**, 117-126.

[621] Shibata, R. (1977). Convergence of least squares estimates of autoregressive parameters, *Austral. J. Statist.*, **19**, 226-235.

[622] Shibata, R. (1980). Asymptotically efficient selection of the order of the model for estimating parameters of a linear process, *Ann. Statist.*, **8**, 147-164.

[623] Shibata, R. (1981a). An optimal selection of regression variables, *Biometrika*, **68**, 45-54.

[624] Shibata, R. (1981b). An optimal autoregressive spectral estimate, *Ann. Statist.*, **9**, 300-306.

[625] Shibata, R. (1983). A theoretical view of the use of AIC, in *Time Series Analysis: Theory and Practice 4*, O. D. Anderson, Ed., North-Holland, Amsterdam, 237-244.

[626] Shibata, R. (1984). Approximate efficiency of a selection procedure for the number of regression variables, *Biometrika*, **71**, 43-49.

[627] Shibata, R. (1985). Various model selection techniques in time series analysis, in *Handbooks of Statistics, Vol. 5*, E. J. Hannan, P. R. Krishnaiah and M. M. Rao, Eds., North-Holland, Amsterdam, 179-187.

[628] Shibata, R. (1986). Consistency of model selection and parameter estimation, in *Essays in Time Series and Applied Processes*; Special volume 23A of *J. Appl. Prob.*, J. Gani and M. B. Priestley, Eds., The Applied Probability Trust, Sheffield, 127-141.

[629] Silverman, H. F. (1977). Introduction to programming Winograd Fourier transform algorithm (WFTA), *IEEE Trans. Acoustics, Speech, Signal Processing*, **25**, 152-165.

[630] Silvey, D. S. (1959). The Lagrangean multiplier test, *Ann. Math. Statist.*, **30**, 389-407.

[631] Sims, C. A. (1988). Bayesian skepticism on unit root econometrics, *J. Economic Dynamics and Control*, **12**, 463-474.

[632] Sims, C. A., J. H. Stock and M. W. Watson (1990). Inference in linear time series models with some unit roots, *Econometrica*, **58**, 113-144.

[633] Slutzky, E. (1927). The summation of random causes as the source of cyclic processes, (Russian with an English summary), *Problems of Economic Conditions*, **3**, No. 1, The Conjuncture Institute, Moscow. A revised English version (1937) appeared in *Econometrica*, **5**, 105-146.

[634] Smylie, D. G., G. K. C. Clarke and T. J. Ulrych (1973). Analysis of irregurarities in the earth's rotation, in *Methods in Computational Physics*, **13**, 391-430.

[635] Solo, V. (1984). The order of differencing in ARIMA models, *J. Amer. Statist. Assoc.*, **79**, 916-921.

[636] Solo, V. (1986a). Identifiability of time series models with errors in variables, in *Essays in Time Series and Applied Processes*; Special volume 23A of *J. Appl. Prob.*, J. Gani and M. B. Priestley, Eds., The Applied Probability Trust, Sheffield, 63-71.

[637] Solo, V. (1986b). *Topics in Advanced Time Series Analysis*, Lecture Notes in Mathematics, Springer-Verlag, New York.

[638] Spliid, H. (1983). A fast estimation method for the vector autoregressive moving average model with exogeneous variables, *J. Amer. Statist. Assoc.*, **78**, 843-849.

[639] Stigum, B. P. (1974). Asymptotic properties of dynamic stochastic parameter estimates (III), *J. Multivariate Anal.*, **4**, 351-381.

[640] Stine, R. A. and P. Shaman (1990). Bias of autoregressive spectral estimators, *J. Amer. Statist. Assoc.*, **85**, 1091-1098.

[641] Stoica, P. (1979). Comments on paper by R. L. Kashap, *IEEE Trans. Automatic Control*, **AC-24**, 516-518.

[642] Stoica, P., P. Eykhoff, P. Janssen and T. Söderström (1986). Model-structure selection by cross-validation, *Inter. J. Control*, **43**, 1841-1878.

[643] Stoica, P. and A. Nehorai (1986). The poles of symmetric linear prediction models lie on the unit circle, *IEEE Trans. Acoustics, Speech, Signal Processing*, **ASSP-34**, 1344-1346.

[644] Stoica, P. and A. Nehorai (1987). On stability and root location of linear prediction models, *IEEE Trans. Acoustics, Speech, Signal Processing*, **ASSP-35**, 582-584.

[645] Stoica, P. and A. Nehorai (1988). On linear prediction models constrained to have unit-modulus poles and their use for sinusoidal frequency estimation, *IEEE Trans. Acoustics, Speech, Signal Processing*, **ASSP-36**, 940-942.

[646] Stoica, P., T. Söderström, A. Ahlén and G. Solbrand (1984). On the asymptotic accuracy of pseudo-linear regression algorithms, *Inter. J. Control*, **39**, 115-126.

[647] Stoica, P., T. Söderström, A. Ahlén and G. Solbrand (1985). On the convergence of pseudo-linear regression algorithms, *Inter. J. Control*, **41**, 1429-1444.

[648] Stone, M. (1974). Cross-validatory choice and assessment of statistical predictions, *J. Roy. Statist. Soc. Ser. B*, **36**, 111-147.

[649] Stone, M. (1977). An asymptotic equivalence of choice of model by cross-validation and Akaike's criterion, *J. Roy. Statist. Soc. Ser. B*, **39**, 44-47.

[650] Stone, M. (1979). Comments on model selection criteria of Akaike and Schwarz, *J. Roy. Statist. Soc. Ser. B*, **41**, 276-278.

[651] Strobach, P. (1990). *Linear Prediction Theory*, Springer-Verlag, Berlin.

[652] Subba Rao, T. and M. M. Gabr (1989). The estimation of spectrum, inverse spectrum and autocovariances of a stationary time series, *J. Time Series Anal.*, **10**, 183-202.

[653] Sugiura, N. (1978). Further analysis of the data by Akaike's information criterion and the finite corrections, *Commun. Statist.*, **A7**, 13-26.

[654] Swift, A. L. (1990). Orders and initial values of non-stationary multivariate ARMA models, *J. Time Series Anal.*, **11**, 349-359.

[655] Takemura, A. (1984). A generalization of autocorrelation and partial autocorrelation function useful for identification of ARMA(p, q) process, Technical Report No.11, Department of Statistics, Stanford University, Stanford.

[656] Tanaka, K. (1983). The one-sided Lagrange multiplier test of the AR(p) model *vs.* the AR(p) model with measurement error, *J. Roy. Statist. Soc. Ser. B*, **45**, 77-80.

[657] Tanaka, K. (1984). An asymptotic expansion associated with the maximum likelihood estimators in ARMA models, *J. Roy. Statist. Soc. Ser. B*, **46**, 58-67.

[658] Tanaka, K. (1986). Asymptotic expansions for time series, in *Essays in Time Series and Applied Processes*; Special volume 23A of *J. Appl. Prob.*, J. Gani and M. B. Priestley, Eds., The Applied Probability Trust, Sheffield, 211-227.

[659] Taniguchi, M. (1980). On the selection of the order of the spectral density model of a stationary process, *Ann. Inst. Statist. Math.*, **32**, 401-409.

[660] Taniguchi, M. (1985). An asymptotic expansion for the distribution of the likelihood ratio criterion for a Gaussian autoregressive moving average under a local alternative, *Economic Theory*, **1**, 73-84.

[661] Taniguchi, M. (1988). Asymptotic expansions of the distributions of some test statistics for Gaussian ARMA processes, *J. Multivariate Anal.*, **27**, 494-511.

[662] Taylor, C. C. (1987). Akaike's information criterion and the histogram, *Biometrika*, **74**, 636-639.

[663] Thomson, D. J. (1981). Some recent developments in spectrum and harmonic analysis, *Computer Sciences and Statistics: Proc. 13th Symposium on the Interface*, 167-171.

[664] This week's citation classic (1981). *Current Contents; Engineering, Technology and Applied Sciences*, **No. 51**, 22, December 21, 1981.

[665] Tiao, G. C. (1985). Autoregressive moving average models, intervention problems and outlier detection in time series, in *Handbooks of Statistics, Vol. 5*, E. J. Hannan, P. R. Krishnaiah and M. M. Rao, Eds., North-Holland, Amsterdam, 85-118.

[666] Tiao, G. C. and G. E. P. Box (1981). Modelling multiple time series with applications, *J. Amer. Statist. Assoc.*, **76**, 802-816.

[667] Tiao, G. C. and R. S. Tsay (1983a). Consistent properties of least squares estimates of autoregressive parameters in ARMA models, *Ann. Statist.*, **11**, 856-871.

[668] Tiao, G. C. and R. S. Tsay (1983b). Multiple time series modeling and extended sample cross-correlation, *J. Busi. Econ. Statist.*, **1**, 43-56.

[669] Tiao, G. C. and R. S. Tsay (1989). Model specification in multivariate time series, *J. Roy. Statist. Soc. Ser. B*, **51**, 157-213.

[670] Tjøstheim, D. and J. Paulsen (1982). Empirical identification of multiple time series, *J. Time Series Anal.*, **3**, 265-282.

[671] Tjøstheim, D. and J. Paulsen (1983). Bias of some commonly used time series estimates, *Biometrika*, **70**, 389-399.

[672] Tjøstheim, D. and J. Paulsen (1985). Least squares estimates and order determination procedures for autoregressive processes with a time dependent variance, *J. Time Series Anal.*, **6**, 117-133.

[673] Tong, H. (1975a). Autoregressive model fitting with noisy data by Akaike's information theory, *IEEE Trans. Information Theory*, **IT-21**, 476-480.

[674] Tong, H. (1975b). Determination of the order of a Markov chain by Akaike's information criterion, *J. Appl. Prob.*, **12**, 488-497.

[675] Tong, H. (1976). Fitting a smooth moving average to noise data, *IEEE Trans. Information Theory*, **IT-22**, 493-496.

[676] Tong, H. (1977). More on autoregressive model fitting with noisy data by Akaike's information criterion, *IEEE Trans. Information Theory*, **IT-23**, 409-410.

[677] Tong, H. (1979). A note on a local equivalence of two recent approaches to autoregressive order determination, *Inter. J. Control*, **29**, 441-446.

[678] Tong, H. (1988). A note on local parameter orthogonality and Levinson-Durbin algorithm, *Biometrika*, **75**, 788-789.

[679] Toyooka, Y. (1982). Prediction error in a linear model with estimated parameters, *Biometrika*, **69**, 453-459.

[680] Trench, W. F. (1964). An algorithm for the inversion of finite Toeplitz matrices, *SIAM J.*, **12**, 515-522.

[681] Trench, W. F. (1974). Inversion of Toeplitz band matrices, *Math. Comp.*, **28**, 1089-1095.

[682] Trench, W. F. (1989). Numerical solution of the eigenvalue problem for Hermitian Toeplitz matrices, *SIAM J. Matrix Anal. Appl.*, **10**, 135-146.

[683] Tsay, R. S. (1989). Identifying multivariate time series models, *J. Time Series Anal.*, **10**, 357-372.

[684] Tsay, R. S. and G. C. Tiao (1984). Consistent estimates of autoregressive parameters and extended sample autocorrelation function for stationary and nonstationary ARMA models, *J. Amer. Statist. Assoc.*, **79**, 84-96.

[685] Tsay, R. S. and G. C. Tiao (1985). Use of canonical analysis in time series model identification, *Biometrika*, **72**, 299-315.

[686] Tsay, R. S. and G. C. Tiao (1990). Asymptotic properties of multivariate nonstationary processes with applications to autoregressions, *Ann. Statist.*, **18**, 220-250.

[687] Tucker, W. T. (1982). On the Padé table and its relationship to the R and S arrays and ARMA modeling, *Commun. Statist.*, **A11**, 1335-1379.

[688] Tunnicliffe-Wilson, G. (1973). The estimation of parameters in multivariate time series models, *J. Roy. Statist. Soc. Ser. B*, **35**, 76-85.

[689] Ulrych, T. J. and T. N. Bishop (1975). Maximum entropy spectral analysis and autoregressive decomposition, *Review of Geophysics and Space Physics*, **13**, 183-200. Reprinted in *Modern Spectrum Analysis*, D. G. Childers, Ed., IEEE Press, New York, 54-71.

[690] Ulrych, T. J. and R. W. Clayton (1976). Time series modeling and maximum entropy, *Phys. Earth Planetary Interiors*, **12**, 188-200.

[691] Ulrych, T. J. and M. Ooe (1979). Autoregressive and mixed autoregressive-moving average models and spectra, in *Nonlinear Methods of Spectral Analysis*, S. Haykin, Ed., Springer-Verlag, New York, 73-126.

[692] Unbehauen, H. and B. Göhring (1974). Tests for determining model order in parameter estimation, *Automatica*, **10**, 233-244.

[693] Van Campenhout, J. M. and T. M. Cover (1981). Maximum entropy and conditional probability, *IEEE Trans: Information Theory*, **IT-27**, 483-489.

[694] Van den Boom, A. J. W. and A. W. M. Van den Enden (1974). The determination of the orders of process and dynamics, *Automatica*, **10**, 245-256.

[695] Van den Bos, A. (1971). Alternative interpletation of maximum entropy spectral analysis, *IEEE Trans. Information Theory*, **IT-17**, 493-494. Reprinted in *Mordern Spectrum Analysis*, D. G. Childers, Ed., IEEE Press, New York, 73-126.

[696] Vaninskii, K. L. and A. M. Yaglom (1990). Stationary processes with a finite number of non-zero canonical correlations between future and past, *J. Time Series Anal.*, **11**, 361-375.

[697] Vasicek, O. A. (1980). A conditional law of large numbers, *Ann. Prob.*, **8**, 142-147.

[698] Velu, R. P., G. C. Reinsel and D. W. Wichern (1986). Reduced rank models for multiple time series, *Biometrika*, **73**, 105-118.

[699] Velu, R. P., D. W. Wichern and G. C. Reinsel (1987). A note on non-stationarity and canonical analysis of multiple time series models, *J. Time Series Anal.*, **8**, 479-487.

[700] Veres, S. (1987). Asymptotic distributions of likelihood ratios for overparameterized ARMA processes, *J. Time Series Anal.*, **8**, 345-357.

[701] Vincze, I. (1982). On the maximum probability principle in statistical physics, *Progress in Statistics: Colloquia Mathematico Societatis Janos Bolyai*, **9-II**, 869-893.

[702] Wahlberg, B. (1989a). Estimation of autoregressive moving-average models via high-order autoregressive approximations, *J. Time Series Anal.*, **10**, 283-299.

[703] Wahlberg, B. (1989b). Model reductions of high-order estimated models: the asymptotic ML approach, *Inter. J. Control*, **49**, 169-192.

[704] Wald, A. (1943). Test of statistical hypotheses concerning several parameters when the number of observations is large, *Trans. Amer. Math. Soc.*, **54**, 426-482.

[705] Walker, A. M. (1950). Note on a generalization of the large sample goodness of fit test for linear autoregressive schemes, *J. Roy. Statist. Soc. Ser. B*, **12**, 102-107.

[706] Walker, A. M. (1952). Some properties of the asymptotic power functions of goodness-of-fit tests for linear autoregressive schemes, *J. Roy. Statist. Soc. Ser. B*, **14**, 117-134.

[707] Walker, A. M. (1962). Large sample estimation of parameters for autoregressive processes with moving average residuals, *Biometrika*, **49**, 117-131.

[708] Walker, A. M. (1964). Asymptotic properties of least-squares estimates of parameters of the spectrum of a stationary non-deterministic time-series, *J. Australian Mathematical Soc.*, **4**, 363-384.

[709] Watson, G. A. (1973). An algorithm for the inversion of block matrices of Toeplitz form, *J. Assoc. Comput. Mach.*, **20**, 409-415.

[710] Wegman, E. J. and J. G. Smith, Eds. (1984). *Statistical Signal Processing*, Marcel Dekker, New York.

[711] Wellstead, P. E. (1978). An instrumental product moment test for model order estimation, *Automatica*, **14**, 89-91.

[712] White, J. S. (1961). Asymptotic expansions for the mean and variance of the serial correlation coefficient, *Biometrika*, **48**, 85-94.

[713] Whittle, P. (1951). *Hypothesis Testing in Time Series Analysis*, Ph.D. Thesis, Uppsala University, Uppsala.

[714] Whittle, P. (1952a). Tests of fit in time series, *Biometrika*, **39**, 309-318.

[715] Whittle, P. (1952b). The simultaneous estimation of a time series harmonic components and covariance structure, *Trabajos. Estadist.*, **3**, 43-57.

[716] Whittle, P. (1953a). The analysis of multiple stationary time series, *J. Roy. Statist. Soc. Ser. B.*, **15**, 125-139.

[717] Whittle, P. (1953b). Estimation and information in stationary time series, *Arkiv för Mathematik*, **2**, 423-434.

[718] Whittle, P. (1954). Some recent contributions to the theory of stationary processes, in *A Study in the Analysis of Stationary Time Series (2nd Edition)* by H. Wold, Almqvist and Wiksells, Uppsala.

[719] Whittle, P. (1962). Gaussian estimation in stationary time series, *Bull. Inter. Statist. Inst.*, **39**, 105-129.

[720] Whittle, P. (1963). On the fitting of multivariate autoregressions, and the approximate canonical factorization of a spectral density matrix, *Biometrika*, **50**, 129-134.

[721] Wiggins, R. A. and E. A. Robinson (1965). Recursive solution to the multichannel filtering problem, *J. Geophys. Res.*, **70**, 1885-1891.

[722] Wilkes, D. M. and M. H. Hayes (1987). An eigenvalue recursion for Toeplitz matrices, *IEEE Trans. Acoustics, Speech, Signal Processing*, **ASSP-35**, 907-909.

[723] Wilks, S. S. (1932). Certain generalization in the analysis of variance, *Biometrika*, **24**, 471-494.

[724] Wilks, S. S. (1938). The large sample distribution of the likelihood ratio for testing composite hypotheses, *Ann. Math. Statist.*, **9**, 60-62.

[725] Wilson, G. (1969). Factorization of the covariance generating function of a pure moving average process, *SIAM J. Numer. Anal.*, **6**, 1-7.

[726] Wincek, M. A. and G. C. Reinsel (1986). An exact maximum likelihood estimation procedure for regression-ARMA time series models with possibly nonconsecutive data, *J. Roy. Statist. Soc. Ser. B*, **48**, 303-313.

[727] Wise, J. (1956). Stationarity conditions for stochastic processes of the autoregressive and moving-average type, *Biometrika*, **43**, 215-219.

[728] Wold, H. (1938). *A Study in the Analysis of Stationary Time Series*, Almqvist and Wiksells, Uppsala.

[729] Wold, H. (1949). A large-sample test for moving averages, *J. Roy. Statist. Soc. Ser. B*, **11**, 297-305.

[730] Wold, H. O. A. (1966). *Bibliography on Time Series and Stochastic Processes*, Oliver and Boyd Ltd., London.

[731] Woodroofe, M. (1982). On model selection and the Arc sine laws, *Ann. Statist.*, **10**, 1182-1194.

[732] Woodside, C. M. (1971). Estimation of the order of linear systems, *Automatica*, **7**, 727-733.

[733] Woodward, W. A. and H. L. Gray (1981). On the relationship between the S array and the Box-Jenkins method of ARMA model identification, *J. Amer. Statist. Assoc.*, **76**, 579-587.

[734] Yajima, Y. (1985). Asymptotic properties of the sample autocorrelations and partial autocorrelations of a multiplicative ARIMA process, *J. Time Series Anal.*, **6**, 187-201.

[735] Yamamoto, T. (1976). Asymptotic mean square prediction error for an autoregressive model with estimated coefficients, *Appl. Statist.*, **25**, 123-127.

[736] Yamamoto, T. (1981). Predictions of multivariate autoregressive-moving models, *Biometrika*, **68**, 485-492.

[737] Yamamoto, T. and N. Kunitomo (1984). Asymptotic bias of the least squares estimator for multivariate autoregressive models, *Ann. Inst. Statist. Math.*, **36**, Part A, 419-430.

[738] Young, P. C. (1974). Recursive approaches to time series analysis, *Bulletin of the Institute of Mathematics and its Applications*, **10**, 209-224.

[739] Young, P. C. (1984). *Recursive Estimation and Time-Series Analysis: An Introduction*, Springer-Verlag, Berlin.

[740] Young, P. C. (1985). Recursive identification, estimation and control, in *Handbooks of Statistics, Vol. 5*, E. J. Hannan, P. R. Krishnaiah and M. M. Rao, Eds., North-Holland, Amsterdam, 213-255.

[741] Young, P. C. and A. Jakeman (1979). Refined instrumental variable methods of recursive time-series analysis, Part I. single input, single output systems, *Inter. J. Control*, **29**, 1-30.

[742] Young, P. C., A. Jakeman and R. McMurtrie (1980). An instrumental variable method for model order identification, *Automatica*, **16**, 281-294.

[743] Yule, G. U. (1921). On the time-correlation problem with especial reference to the variate-difference correlation method, *J. Roy. Statist. Soc.*, **84**, 497-537.

[744] Yule, U. (1927). On a method of investigating periodicities in disturbed series, with special reference to Wolfer's sunspot numbers, *Philosophical Transactions of the Royal Society of London, Ser. A*, **226**, 267-298.

[745] Zellner, A., Ed. (1978) *Seasonal Analysis of Economic Time Series*, U.S. department of Commerce, Bureau of the Census, Washington, DC.

[746] Zhang, X. D. and H. Takeda (1987). An approach to time series analysis and ARMA spectral estimation, *IEEE Trans. Acoustics, Speech, Signal Processing*, **ASSP-35**, 1303-1313.

[747] Zohar, S. (1969). Toeplitz matrix inversion: The algorithm of W. F. Trench, *J. Assoc. Comput. Mach.*, **16**, 592-601.

[748] Zohar, S. (1974). The solution of a Toeplitz set of linear equations, *J. Assoc. Comput. Mach.*, **21**, 272-276.

[749] Zohar, S. (1979). FORTRAN subroutines for the solution of Toeplitz sets of linear equations, *IEEE Trans. Acoustics, Speech, Signal Processing*, **ASSP-27**, 656-658.

Index

Springer Series in Statistics

(continued from p. ii)

Shedler: Regeneration and Networks of Queues.
Siegmund: Sequential Analysis: Tests and Confidence Intervals.
Todorovic: An Introduction to Stochastic Processes and Their Applications.
Tong: The Multivariate Normal Distribution.
Vapnik: Estimation of Dependences Based on Empirical Data.
West/Harrison: Bayesian Forecasting and Dynamic Models.
Wolter: Introduction to Variance Estimation.
Yaglom: Correlation Theory of Stationary and Related Random Functions I: Basic Results.
Yaglom: Correlation Theory of Stationary and Related Random Functions II: Supplementary Notes and References.